System Theory and Artificial System Design

系统理论与人工系统设计学

Wang Yue   Jia Lijuan
王　越　　贾丽娟 ◎ 著

版权专有　侵权必究

图书在版编目（ＣＩＰ）数据

系统理论与人工系统设计学 ＝ System Theory and Artificial System Design：英文／王越，贾丽娟著．－－北京：北京理工大学出版社，2023.5
　　ISBN 978-7-5763-2387-0

　Ⅰ．①系⋯　Ⅱ．①王⋯ ②贾⋯　Ⅲ．①系统理论-英文 ②系统设计-英文　Ⅳ．①N941 ②N945.23

中国国家版本馆 CIP 数据核字（2023）第 086371 号

出版发行 ／	北京理工大学出版社有限责任公司
社　　址 ／	北京市海淀区中关村南大街 5 号
邮　　编 ／	100081
电　　话 ／	（010）68914775（总编室）
	（010）82562903（教材售后服务热线）
	（010）68944723（其他图书服务热线）
网　　址 ／	http://www.bitpress.com.cn
经　　销 ／	全国各地新华书店
印　　刷 ／	三河市华骏印务包装有限公司
开　　本 ／	710 毫米×1000 毫米　1/16
印　　张 ／	14.25
字　　数 ／	267 千字
版　　次 ／	2023 年 5 月第 1 版　2023 年 5 月第 1 次印刷
定　　价 ／	89.00 元

责任编辑 ／	徐　宁
文案编辑 ／	把明宇
责任校对 ／	刘亚男
责任印制 ／	李志强

图书出现印装质量问题，请拨打售后服务热线，本社负责调换

Preface

Human society is a whole and a representation of the survival and development of all human beings, which is extremely important to human beings but cannot be fully known and mastered. Mr. Qian Xuesen innovatively proposed that human society is an "open complex giant system", which reflects the continuous evolutionary development of human society from low to high levels and from simple to complex, and the cognitive and practical abilities of human beings are also integrated into the evolutionary development process of human society from low to high levels. The introduction and application of the concept of "system" is a qualitative change in human understanding and practice. Artificial systems are systems of platforms and tools designed and built by humans to serve humans, and are integral to the advanced evolutionary development of human understanding and practice. Artificial systems design is an evolving systemic science for the development and design of artificial systems. This book focuses on the design of artificial systems in the field of information and is a summary of the author's many years of experience in the field of information research. The idea behind this book is to address the problems faced by systems theory and artificial systems design, and to implement the top-level philosophy and theory of systems into concrete applications through the proposed theoretical approach. Therefore, in the important spirit of "promoting the development of a community with a shared future for mankind, everyone has the responsibility", the author has named this book *System Theory and Artificial System Design*, hoping to contribute to the development of human society and the development of system theory science.

The book is divided into seven chapters. Using materialistic dialectics as a top-level ideological and theoretical guide, the book describes the current state of development of systems theory using dissipative self-organization theory, self-organizing evolutionary theory, and open complex giant systems theory as theoretical foundations. This is followed by a detailed discussion of information and information systems, the recognition, representation and description of systems, and the development, classification and design philosophy of artificial systems. The main points of the thinking, methods and processes of artificial system design are summarized. A method for applying systems theory to practical design, the multi-living agent theory approach, is also presented. Among them, as the applied foundational theories are generally general and universal, more abstract and difficult to understand, they are mostly combined with examples in the course of the exposition.

The first chapter is an overview. It discusses the basic laws of the interrelated and mutually supportive co-evolution of philosophy and natural science, systems science, and artificial systems design and dialectical materialistic philosophy, laying the foundation for the subsequent chapters of the book to combine the application of the law of dialectical contradiction and unity of opposites. Chapter 2 deals with the foundations of systems theory. A definition of the 'system' based on dissipative self-organization theory is given, together with a description of its main features. The categories of opposites and unity involved in systems theory are given. The axiomatic system of systems is given. The key entry point for the study of systems is given - "relations". Chapter 3 covers the development situation of systems theory. The development of systems theory as a system of development is briefly described in terms of the development of dissipative self-organization theory, self-organizing evolutionary theory, and open complex giant systems theory. Chapter 4 deals with information and information systems. It deals with theoretical and technical issues related to information, information systems, and information technology, and clarifies that information technology and information systems are one of the foundations of the development of the information society. Chapter 5 deals with the recognition, characterization, and description of systems. It introduces the basic concepts and methods of system recognition, characterization, and description, and focuses on the author's theoretical approach of "multi-living agents", which can be applied to support the human-led implementation of artificial systems with better service functions and longer service cycles by strengthening the relevant applications. Chapter 6 is on the development, classification, and design philosophy of artificial systems. Artificial system characteristics, classification, and development elements are discussed to promote artificial systems to better serve human development; and artificial system design philosophy is discussed to enhance the success rate of complex artificial system design. Chapter 7 is the main nodes of thinking, methods, and processes for the design of artificial systems. The main theme is "Artificial System Design", and the design process is summarized comprehensively in the context of practical development, resulting in several reference frameworks such as: problem-solving thinking, system-level design thinking, a cuttable four-dimensional indicator system, guidelines to be followed in design, and the main steps in design.

This book addresses the problems facing systems theory and artificial systems design science, based on the problem of artificial systems design in the field of information and information systems, based on the development of existing systems theory, the theory, methods, and practice of artificial systems design are systematically addressed. It also briefly introduces some methods of recognizing the dialectical way of thinking in resolving contradictions. The corresponding ideas and methods are provided for relevant scholars to conduct their research. It is important to note that this is a developing and highly complex discipline, with much uncharted territory and much to explore, awaiting the diligent exploration and intelligent creation of many, many aspirants. The potential and the results will be fascinating. At the same time, the authors argue that in the process of exploring innovation, there is a need to closely integrate applied fundamental and applied-level research in order to strengthen the drive for innovative development.

The author believes that in the exploration and innovation journey to achieve the great rejuvenation of the Chinese nation, it is necessary to continuously resolve conflicts scientifically and efficiently, and it is necessary to closely combine the research of the application foundation layer and the application layer to strengthen the "driving force of innovation and development", so as to achieve the goal of "strengthening the country in science and technology".

Contents

Chapter 1　Overview ·· 1

  1.1　Introduction ·· 1
    1.1.1　The Inevitable Trend of the Development of System Theory ··············· 1
    1.1.2　Artificial System Design ·· 2
  1.2　Relationship Between System Theory and Philosophy as
       well as Other Relevant Disciplines ··· 3
    1.2.1　Philosophy ··· 3
    1.2.2　Relationship Between Philosophy and Natural Sciences ···················· 9
    1.2.3　Relationship Between System Theory and Philosophy ····················· 11
    1.2.4　Relationship Between System Theory and Other
          Disciplines (Natural Sciences) ·· 12

Chapter 2　Basics of System Theory ·· 15

  2.1　Definition of System and Explanations of Its Key Points ························· 15
    2.1.1　Definition of System ··· 15
    2.1.2　Explanations of Its Key Points ··· 15
  2.2　Categories of the Unity of Opposites Often Concerned with the
       Application of System Theory ··· 16
    2.2.1　Relativity and Absoluteness ··· 17
    2.2.2　Similarity (Resemblance) and Difference (the Comparison
          Between the Characteristics of Things) ·· 18
    2.2.3　Certainty and Contingency (the Representation of the Compliance
          of the Motion of Things with Laws Mastered by Mankind) ············· 18
    2.2.4　Purposefulness and Natural Determinism (the Representation
          of the Compatibility of Man's Intentions with Objective Laws
          and Conditions) ·· 19
    2.2.5　Whole and Parts (the Relationship Between Holistic and Partial
          Existence in Things) ·· 20
    2.2.6　Complexity and Simplicity (General Characteristics of Mankind's
          Knowledge and Assessment of the Motion and Existence of Things) ··· 20

2.2.7 Qualitative and Quantitative Changes (Representing a Pair of the Most Basic Motion Characteristics in the Motion of Things) 21

2.2.8 Continuous and Discontinuous (the Representation of the Continuation of the Changes, the Adjacency of Certain Important Characteristics in the Motion of Things in Temporal and Spatial Order) 21

2.3 Unavoidable Brief Discussions on "Beauty" 22

2.3.1 The Essence of Men and Beauty 22

2.3.2 The Categorization of Beauty 23

2.3.3 Scientific Beauty 23

2.3.4 Technical Beauty 25

2.3.5 Engineering Beauty 25

2.3.6 Scientific and Engineering Beauty in the Fields of System 26

2.4 Axioms of System 26

2.4.1 Top Axioms of System Theory ("Shared" with Materialist Dialectics) 27

2.4.2 Temporary System of Axioms of System Theory 27

2.5 The Expanded Model of Multilayered Temporal and Spatial Domains of Systematic Motion 29

2.5.1 A Basic Theoretical Framework Behind the Establishment of This Model is Composed of the Following Core "Elements" 30

2.5.2 Multilayered Multi-periodical Expansion Model of System Based on Relations (Represented Separately in Terms of Space and Time) 30

2.6 Relations 34

2.6.1 Mathematic Definitions 34

2.6.2 Connotations of Relation in System Theory 35

2.6.3 Some Important Relations Concerned with in System Theory 36

2.6.4 The Classification of "Relations" from the Perspective of the Connotations and Denotations of "the Concept of Relation" 37

2.6.5 Relations of the Objects Sustained by "Relations" (Set Objects as XYZ) 37

2.6.6 Spatial and Temporal Expansion of Relations 38

2.7 Ideal Model of System and Its Application 38
2.8 Thoughts on the "Improvement" of the Efficiency of Computers 40

Chapter 3 The Current Development of System Theory 41

3.1 Dissipative Structure—Self-Organizing System Theory and Its Systematic Framework 41
3.2 Dissipative Structure-Self-Organization Theory 42
    3.2.1 Dissipative Theory Proposed by Professor Prigogine and the Theoretical Basics of the Generation of Minimum Entropy 42
    3.2.2 The Principle of the Generation of New Order When Far Away from the Equilibrium 43
    3.2.3 Kinetic Process Model (Blusseletor) 44
3.3 Self-Organizing Evolution—Paradigm of Self-Organization 45
    3.3.1 Dissipative Structure: The Necessary Basic Structural Mode of the Survival and Evolutionary System 45
    3.3.2 Macro-Order of a System 46
    3.3.3 Through Fluctuations the System Reaches a New Order: the Beginning of the New-Level and Process of the Evolution of the System 46
    3.3.4 The Co-Evolution of Macro-and Micro-Cosmos: the Common Evolution of the Environment and the System Itself 48
    3.3.5 The Evolution of the Evolutionary Process (More Applicable for the Evolution of Human Society) 53
3.4 Professor Qian Xuesen's Propositions and Contributions 54
    3.4.1 The Proposition of the Concept of "Open Complex Giant System" 54
    3.4.2 From Qualitative to Quantitative Synthesis—the Core Methods of Research on Giant Complex Open Systems 55

Chapter 4 Information and Information Systems 59

4.1 Information 59
    4.1.1 Connotations and Definition of Information 59
    4.1.2 Representation and Characteristics of Information 60
    4.1.3 "Information's" Characteristics and Summary of Its Characteristics Concerned 61
    4.1.4 The Progress of Mankind Transferring and Utilizing Information 67
4.2 Information Systems 68

  4.2.1 Definition of Information Systems ················· 68
  4.2.2 The Systematic Theoretic Characteristics of "Information Systems" ······ 69
  4.2.3 Functional Constitution of Information Systems ················· 69
  4.2.4 Basics of the Development of Information Systems ················· 72
  4.2.5 The Ultimate Limit Goal of the Development of Information Systems ··· 76
 4.3 Development of Information Science and Technology and Information
   Systems: One of the Eternal Themes for Human Beings ················· 77
  4.3.1 Theoretical Discussions ················· 77
  4.3.2 Information Science and Technology and Systems as Universal
    "Enhancer" and "Catalyst" ················· 77
  4.3.3 Information Systems Offer Services in a Wide Range ················· 78
  4.3.4 Exemplifications of Electronic and Optoelectronic Information Systems
    in Vehicles (Huge Series of Products and Industry Groups Are Already
    Formed) ················· 80
 4.4 Huge Support Systems for the Development of Information Science
   and Technology and Information Systems ················· 80
  4.4.1 Discipline Backup Systems ················· 81
  4.4.2 Social Supports and Major Institutions Undertaking Tasks Necessary
    for the Development of Different Levels ················· 81
 4.5 Examples of Typical Information Systems and Their Main Characteristics ··· 82
  4.5.1 National Defense Information Systems ················· 82
  4.5.2 Recent Developments in Information Systems for Civilian Use ········· 83
  4.5.3 Discussions on Some Basic Concepts of High-Tech Enterprises
    in the Field of Information ················· 85
 4.6 Conclusion of this Chapter ················· 86

Chapter 5 Basics of System Knowledge, Representation and Description ··· 88

 5.1 Introduction ················· 88
 5.2 Basic Concepts of System Description and Representation ················· 88
  5.2.1 Proper Understanding of "Human-Centered" Concept ················· 88
  5.2.2 The Concept of "Relations" ················· 89
  5.2.3 The Concept of "Mapping" ················· 90
  5.2.4 The Concept of Operation ················· 90
  5.2.5 The Concept of Generalization, Extraction, and Abstract Modeling ··· 90
 5.3 Preliminary Study on Representation and Description of System ············ 91

Contents

- 5.3.1 General Overview of System Representation and Description Fundamentals 91
- 5.3.2 Basic Thinking Modes in System Representation and Description 92
- 5.3.3 Brief to Some Commonly Used Basic Methods 93
- 5.3.4 New Field Methods in System Design, Representation and Studies on Complex Scientific Problems and Development of Technical Innovations Based on the Guidance and Support of Computation Science 95
- 5.3.5 The Extension in Applications of System Representation and Description 98
- 5.4 The Theory and Method of Living Self-Organization Mechanism, and the "Multi-Living Agent" 99
  - 5.4.1 Brief Introduction 99
  - 5.4.2 "Livelihood" and Agent Livelihood 100
  - 5.4.3 Expressions of "Agent" in Dynamics (Function State, and Environmental Constraints, etc.) 102
  - 5.4.4 The Living Self-Organization Mechanism (LSOM) and the Expression of Two Set Models 103
  - 5.4.5 The Composition of Multi-Living Agent of System and the Three-Set Model 106
  - 5.4.6 Quantitative Expression of the Livelihood of "Agent" 107
  - 5.4.7 Adjustment and Maintenance of the Livelihood of the Function of the Agent and the Multi-Agent System 111
  - 5.4.8 The Multi-Layer Realization of the Adjustment and Maintenance of the Livelihood of the System Functions of the Multi-Living Agent 118
  - 5.4.9 Negotiation and Coordination of the Adjustment and Maintenance of the Livelihood of the Functions Among Multiple Agents 118
  - 5.4.10 Management and Control Agent 120
  - 5.4.11 A Theoretical Comparison Between the Dissipative Self-Organization Theory (Key Content) and the MLAM Theory 124
  - 5.4.12 Exemplifications of Applications 126

Chapter 6 Development and Categorization of Artificial System Design and the Philosophy of Artificial System Design 137

- 6.1 Discussions on the Essentials of Artificial System 137

6.1.1 Discussions on the Connotations of Artificial System 137
6.1.2 Discussions on the Concise Categorization of Artificial System from the Perspective of Functions and Characteristics 139
6.2 Origins of the Development of Artificial System: Discussions on the Thinking Pattern of Mankind 143
6.2.1 The Essence of Mankind Lies in Their Learning and Practicing Abilities, Which Are Also the Fundamental Causes for the Development of Artificial Systems 143
6.2.2 The Core Elements for the Development of Mankind's Learning and Practical Abilities Are the Development of Thinking Abilities 143
6.2.3 The Scientific Thinking Pattern of Mankind Should Be Dialectical and Follow the Law of Unity of Opposites 143
6.2.4 The Law of the Unity of Opposites is the Basic Law in the Research and Development of Artificial System 144
6.3 Philosophy of the Research and Development of Artificial System Design Under the Guidance of System Theory 145
6.3.1 From Philosophical Discussions to the Philosophy of Artificial System Design 145
6.3.2 Brief Introduction to the Philosophy of Artificial System Design 148

Chapter 7 Major Points in the Design, Thinking, Methods and Process of Artificial Systems 157

7.1 Preface to the Significance of the Expansion of Artificial Systems 157
7.1.1 Major Artificial Systems Are Already an Indispensable Part of Human Society, Whereas an "Innovative Society" is the Scientific Aim and the Core Representation of China's Social Development 157
7.1.2 The Birth of Fine Artificial Systems Involves the Application of Various "Law" Systems, and the Birth of More Fine Artificial Systems and Their Being Incorporated Into Society to Function Are Strongly Indicative of an Innovative Society 158
7.1.3 Innovative Society—the Core Factor in the Harmonious Development of Man and Society is that Human Aptitudes Continually Develop Under the "People-Oriented" Principle 158
7.1.4 Thinking and Thinking Methods 159

Contents

- 7.2 Types of Human Thinking 160
  - 7.2.1 Logical Thinking 160
  - 7.2.2 Figurative Thinking 160
  - 7.2.3 Intuitive Thinking—Function of Human Consciousness Which is of Deeper Level than Figurative Thinking 161
- 7.3 The Transcendental Promotive Thinking Mode (Combination of Methods and Simulation): Following and Using Philosophy to First Impel the Transformation of the Conditions for the Development of Things, Then Impel the Divergence of the Scientific Evolution of Things 161
  - 7.3.1 Philosophical Basis 161
  - 7.3.2 The Transcendental Promotive Thinking Mode, the Core Factor and Move of Practical Application 162
  - 7.3.3 Exemplification 163
- 7.4 Thinking Methodology for the Solution of Problems 163
  - 7.4.1 The Direct Method Conforming to the Consequent Thinking: Grasp the Sensitive Principal Contradictions (in the Current Phase), and Take Measures Following the Direction of Development to Solve the Contradictions Hindering the Development of the Principal Aspects 163
  - 7.4.2 Reasoning Complying with the Direction of Development +Jump Solutions (Reasoning Following the Forward Direction in General Level + Jump Development in Key Technical Level) 164
  - 7.4.3 The Opposite Yet Complementary Methods Using the Reverse Thinking to Solve Contradictions in the Light of Specific Problems, in Which the Important Thing is that the Opposite yet Complementary Effects Can Eventually be Achieved 165
  - 7.4.4 Analogy and Association 167
  - 7.4.5 Transformation 167
  - 7.4.6 Basic Thinking Methodology Used to Solve Specific Problems: Combination of Qualitative and Quantitative Methods (Also Taken as One of the Thinking Modes Solving Specific Problems) 168
  - 7.4.7 Another Thinking Methodology Used to Solve Specific Problems: Combination of Analysis and Synthesis (Also Taken as One of the Thinking Modes Solving Specific Problems) 168
- 7.5 Hierarchical Design Method 169

- 7.5.1 Modeling Analysis and Synthesis Via Human Brain's Abstracted Generalization ... 170
- 7.5.2 Relationship Mapping Inversion (RMI) ... 171
- 7.5.3 Hierarchy Analysis—a Method Combining Quantitative and Qualitative Methods Aiming at Solving the Problem of Decision-Making Under Multi-Factor Environment in the Process of Design ... 173
- 7.5.4 Design Method Supported by Computation Science ... 175
- 7.5.5 Method of Multi-Living Agent ... 176
- 7.6 The Scalable Index System of an Artificial System ... 176
  - 7.6.1 Scalable Four-Dimension Comprehensive Index System ... 176
  - 7.6.2 Connotations of the Index System ... 179
  - 7.6.3 Characteristics of the Quantitative Index of the Index System ... 181
  - 7.6.4 Summary About the Index System ... 182
- 7.7 Design Criteria ... 182
  - 7.7.1 Stage of the Formation of the Objective After the Perception of Contradictions ... 183
  - 7.7.2 Stage of the Formation of the General Core Requirement for a New Artificial System ... 183
  - 7.7.3 Stage of the Formation of Index System ... 183
  - 7.7.4 Stage of the Formation of System Scheme ... 184
  - 7.7.5 Stage of Technical Design ... 184
  - 7.7.6 Stage of the Test Manufacture and Testing of the Model Machine ... 185
- 7.8 Major Steps and Contents of the Design (Development) of Artificial Systems ... 185
  - 7.8.1 Co-Development of Human Society and Artificial Systems ... 185
  - 7.8.2 Major Stages and Contents of Artificial Systems Design (Development) ... 187
  - 7.8.3 Application of the Multi-Living Agent Method to Design ... 190
- 7.9 Perception of the Off-Normal States in Design and the Relevant Handling ... 192
  - 7.9.1 Types of the Off-Normal State Generated Directly in the Design Process and Its Perception ... 192
  - 7.9.2 Perception of the Non-Normal States in a Broad Sense ... 194
  - 7.9.3 General Principles for the Handling of Non-Normal States ... 195

7.9.4　Brief Summary ………………………………………………… 196
7.10　Examples: the Design of Air Terminal Defense Ground Environment
　　　(Based on Certain Foreign System) ……………………………… 196
　　7.10.1　Stage of Formation of the Specific Objective of Solutions Via
　　　　　　Perception of Focal Contradictions ………………………… 196
　　7.10.2　Initial Materialization of the Objective to Form the Core
　　　　　　Requirement of the System …………………………………… 199
　　7.10.3　Formation of the Major Index System ……………………… 205
　　7.10.4　Stage of the Formation of the System Scheme …………… 206
　　7.10.5　Technical Design ……………………………………………… 208
　　7.10.6　Modulations Reflection of Living …………………………… 208

Chapter 1
Overview

1.1 Introduction

1.1.1 The Inevitable Trend of the Development of System Theory

The "world" (which comprises nature and human society) is in ceaseless motion and development. In the 21st Century, with the increasing depth as evinced by the knowledge and practice of mankind, accompanied by the current situation and social demands for development, there is an urgent need for high-level characteristics concerning the motion of things to enhance our knowledge and practice. The essential one of those characteristics is "system", i. e., the comprehension of the motion, development and evolution of all things based on their "systematic" features (the entire dynamics formed by diverse interacting factors). This means that systematic structures and relations between important interactions in the motion of things have to be grasped to achieve all-inclusive and emphatic analysis and understanding of the laws of the motion of things. This is also the inexorable trend of the combination of human beings' capacities and practices accumulated in their evolution and development up to now. It illustrates the leap-forward development of mankind's abilities in their comprehension and solution of complex problems.

The improvement of mankind's knowledge of the "systematic" characteristics of things is the subject of system theory, a developing discipline which is not yet in completion (Things are constantly developing, which entails that they can never reach the state of absolute completion.). Characterized by comprehensive and complex contents, its combinations of quantitative and qualitative features as well as complexity and simplicity in research approaches and methodology are both important and challenging. All those aspects are incorporated into the overall objectives and characteristics of this discipline that studies the laws of the dynamic development of things, instead of adjourning at their existential states. To conclude, system theory is a blooming and thriving subject.

1.1.2 Artificial System Design

Artificial system design is an applied discipline based on system theory, which has wide-ranging uses, meeting the pressing demands of human practices. In those processes, people should make designs and arrangements to the targets from the perspective of the whole system rather than specific dissections and designs, in order to achieve better effects and greater chances of success. Those are the indispensable demands in the process of the development of mankind, which must be satiated.

In addition, this discipline is not perfect. Firstly, it is transcendentally not flawless due to the nonexistence of unadulterated artificial systems. Mostly artificial systems are mixtures with main functions designed artificially (on account that mankind can manufacture neither man nor matter). Secondly, it is difficult in the sense of absolutely correct grasp and application of the rules governing complex system design because of complications and uncertainties in actual applications. To sum up, it is impossible to obtain absolutely accurate certitude concerning the future, in which case there would be no future, no change or no development at all. Thus, precision in design is only a case of relativity, and "perfection" means relative "applicability" in a designed system. Besides, there are contradictions in the "spatial dimension", which refers to the fact that designers and users can be perceived as dispersed points in space, whose mutual relationship is basically characterized by similarity instead of equivalence. Even if reduced to the use demands of the designed products (except for those very simplistic ones), complete equivalences are still impossible to realize. All those factors result in the unfeasibility of absolute success in design. Though the subject is so complicated, it is the inborn nature of human beings to aspire for maximum results. While total perfection is unattainable, comparatively better results that facilitate man's achievement of "freedom" in certain circumstances are still worth our efforts.

"Artificial system design" completely falls into this domain. As a developing and very complex discipline, unknown territories and unexplored subjects still abound, constituting its potential and enchantment, which calls for unfaltering endeavors and intelligent creations of numerous aspirants, as Mao Zedong's saying goes: "we should broaden our vision in the observation of the world".

1.2 Relationship Between System Theory and Philosophy as well as Other Relevant Disciplines

1.2.1 Philosophy

Philosophy is an important ancient discipline, which is nevertheless in constant development. It has diverse secondary disciplines, boasting of many thriving or declining schools as well as historically renowned masters, such as Plato, Aristotle, Kant, Hegel, Marx, Engels, Lenin in the West as well as Confucius, Mencius, Laozi, Zhuangzi, Mozi, Wang Fuzhi, and Cheng Yi in China, who set up inspiring examples for mankind, but also call for inheritance and development. The entire scope of philosophy can be divided into three domains, i.e., Western, Chinese and Indian philosophy. The longtime flourishing of philosophy can be attributed to mankind's need for its common and fundamental nature. Though Western and Chinese philosophies put emphasis on different subjects in discussions, they share common essentials, that is, "theorized worldviews", which are mankind's understanding of the basic laws underlining existence and their "comprehension" of the aspiration for the most common rules governing correct cognitive methods as well as the essence of thinking. This is also the simplest summary of the common and basic nature of philosophy. With the increasing evolution and advancement of human society, knowledge is incessantly approaching the threshold of truth. Philosophical pursuits will never transpire but endlessly develop under the guidance of philosophical "critical" spirit. Actually, the insightful understanding of diverse disciplines in various spheres of knowledge requires the assistance of philosophy and philosophical thinking. Even mathematics, which is a distinguished subject for its abstraction and logic, has need for philosophy!

1.2.1.1 Western Philosophy

Generally speaking, western philosophy has been centered on the cognition, discussion and debate around the basic propositions of "being" and "thinking," carried out from Plato, Aristotle to Kant, Hegel, Marx and Engels as representatives until now.

It is obvious that though the process is packed with conflicts, complications and even retrogressions, the understanding of those basic propositions is in the process of development. In this process, philosophers such as Aristotle, Hegel, Marx, Engels and Lenin are especially worth mentioning.

Aristotle opposed Plato, who was his teacher, for his proposition that "thinking is the first". Instead, he proposed that "being is the first", while "thinking" depends on "being" as the second order. However, he put forward that development is propelled by external forces and didn't realize the internal forces as embodied by oppositions. He is the founder of formal logics, who established the philosophic system which resembles "tao" as explored in Chinese philosophy. Thus, translators drew on the quotation "Therefore what exists before physical form is called the Way" to term his philosophical works "metaphysics" in Chinese (In fact, due to the fact that philosophy is ranked behind physics in his works, it is originally called "Ta meta ta physics," which means "after physics". With the Greek article "ta" omitted, it comes to its Latin and English name "metaphysics".). Because it is excoriated by Hegel for its refusal of dialectics, metaphysics acquires the status of the synonym for anti-dialectics (which is derogative). However, in fact, Aristotle's philosophical thoughts have played a positive role in history. For instance, though he didn't know the dialectical nature of the unity of the opposites, Aristotle proposed many arguments concerning dialectics in his philosophical works, on account of which he was considered by Engels as "the most erudite man" among ancient Greek philosophers (his works touched upon philosophy, psychology, natural sciences, logics, rhetoric, aesthetics, etc.). Engels also thought that he had researched the main form of dialectics and was "Hegel of the ancient world."

In the later period, Hegel emerged as an important western philosopher, who brought German classic idealist philosophy initiated by Kant to consummation, establishing the most extensive objective idealist system in the history of European philosophy. Yet his main contribution lies in his systemization of dialectics, greatly boosting the development of idealist dialectics by elucidating the three dialectic laws: the interaction of qualitative and quantitative changes, the unity of opposites and the negation of negation. He also conjectured the dialectics of objective things per se, and proposed the theory of unity between matter and motion. He has another significant contribution, which is called by Engels as "a great achievement," that is, he is the first philosopher to see the natural, historical and spiritual world as a constantly moving, changing, transforming and developing process. However, the idealist philosophical system and dialectical methodological system established by Hegel are plagued with fundamentally conflicting contradictions, and the revolutionary spirit of his dialectics is suffocated by his idealist system, though his philosophy has acted as one of the theoretical sources for Marxist philosophy.

It is a well known fact that Marx and Engels have established the complete scientific system and framework of dialectical materialist philosophy, which needs no further comment. However, it is worth emphasizing that the common core principle of "the law of the unity of opposites in the development of things" as proposed by dialectical materialism is continuously enriched by the knowledge and practices of mankind, the content of which is not stagnant and changeless but in the process of development. Compared to the era of Marx and Engels, considerable development has been achieved in the content of dialectical materialism in present times. The very value of this philosophical system established by Marx lies in its scientific and developing nature.

Generally speaking, western philosophy treats the relationship between "being" and "knowledge" as its core argument, and the dichotomy of subject and object as its starting point and basis. It stimulates men (subject) to study epistemology on the grounds of commonness and certainty, which further lead to the thriving of science (natural sciences) and the development of material civilization. However, the doctrines of "fixation" and mechanism as proposed in metaphysics have dominated philosophy for a long time, which also lead to absolutism and abstraction in the study of the essence of human, restricting life and individuality. Thus, in the west, there is a saying that laments the phenomenon of "democracy without freedom." Hence the emergence of the postmodernist schools of philosophy, which opposes "the dichotomy of subject and object" as well as "subject theory," bearing similarities to the concept of harmony put forward by Chinese philosophy. I personally think that the law of the unity of opposites in dialectical materialism is absolutely right, and the concept of "unity" should enclose "harmony" of opposites. Though harmony is temporary, comparative and changeable, it is important in certain occasions.

1.2.1.2 Chinese Traditional Philosophy

Chinese philosophy is one of the three most important philosophical systems of the world (Indian philosophy included), differing with western philosophy in fundamental ways, which arises from distinct thinking styles. The three philosophical systems naturally have their own advantages and disadvantages respectively, entailing that it is both wise and necessary for communication and conversation for the purpose of mutual enrichments. Actually, some renowned physicists have expressed their gratitude for Chinese philosophy, which proves to be greatly helpful for their achievements in the field of system science. Chinese thinking style is mainly characterized by generalization, while western thinking style puts emphasis on analysis, particularizing things into tiny details. Though particularization has advantages in seeking for rules and principles, overemphasis on it can lead to tunnel vision and closed-mind of thinking, causing errors. Since the concept of system pays more attention to the wholeness, entirety as well as the evolutionary process, instead of accentuating mere existence and being, Chinese thinking style is worth promulgation in learning complex things.

One of the characteristics of Chinese philosophy is its emphasis on the concept of harmony, advocating "the harmony between Heaven and mankind", with the former referring to nature and the latter certainly pointing to human beings. Harmony in this sense means the coordinated coexistence between men and nature, seeking for (fulfilling) the laws of the unity between nature and mankind. However, it also sees this relationship between nature and men as dynamic, changeable and supplementary, evincing distinctive thoughts that obviously accentuate "internal causes" (internal transcendence). Here are some illustrations. Confucius said that "What has passed will pass, like this (river), going through day and night without pausing."; Laozi put forward that "All things leave behind them the Obscurity (out of which they have come), and go forward to embrace the Brightness (into which they have emerged)."; *I Ching* proposed that "Tao (the Way) means the combination of Yin and Yang" (the process and rules of the change of Tao); Zhu Xi brought forward the concept of "the transformation of Yin and Yang" and commented that "Things will develop in opposite direction when they become extreme."; Laozi observed that "The motion of Tao by contraries proceeds," which means motion is turned towards the opposite direction; Wang Fuzhi argued that: "Harmony is the destiny of Heaven and Earth, as well as the essence of all things," pursuing unity and harmony while emphasizing internal causes for the active role of the subject (internal transcendence), thus putting forward the doctrine of "sage inside, king outside" (the concept of "king outside" is naturally outdated.). (Though not stressing on internal transcendence as much, it has contents that put priority on objectiveness). Actually the concept of "Identifying with the Superior" put forward by Mozi before 400 B. C. has accentuated on objectiveness. Besides, the transcendental thoughts as embodied in Chinese philosophy propose transcending actual concrete "being" to mediate on future development, which is highly abstracted as the transformation between "being with form" and "being without form", as Laozi's saying goes: "Everything on the earth is generated by being with form, and being with form comes from being without form." All of those are the important manifestations of the wisdom of the Chinese nation.("有": being with form; "无": being without form). Since ancient times, Chinese philosophy has already possessed thoughts of simple dialectical materialism, which is a very valuable legacy.

However, with its emphasis on harmony, Chinese philosophy seldom notes the dichotomy between subject and object, which dissects and analyzes things to deepen understanding, as practiced in western philosophy. Meanwhile, too much preference is conferred to generalizing qualitative thinking, while accurate quantitative analysis is devoid. This fact leads to insufficient quantitative analysis that is indispensable for natural sciences, severely obstructing the research and development of China in the fields of natural sciences, which results in the slow advancement of Chinese science and technology after the Middle Ages, and finally brings about her falling behind. The disadvantages of Chinese nation discussed above can be counted as one of the reasons for China's lagging behind. Therefore, in studying and applying Chinese traditional philosophy, when paying attention to "harmony" and "wholeness", we should as well emphasize the resolution and analysis of specific things to compensate for the inadequacy of deepened understanding. In this way, can we realize the organic combination of qualitative and quantitative analysis, part and whole, as well as analysis and generalization for better application.

1.2.1.3 The Combination of Chinese Traditional Philosophy and the Essence of Western Philosophy, that is, Dialectical Materialism, is the Necessary Road for Sustainable Development

Firstly, there is appropriate basis for the combination of Chinese and western philosophies: based on ancient simple materialism and dialectics of Chinese philosophy, through long-term debates, western philosophers established dialectical materialism (though western scholars in the field of philosophy haven't reached a consensus on this point), forming common understandings for development. This is the most important foundation for their combination. Moreover, in history there are many examples showcasing some renowned western philosophers making great contributions by drawing on Chinese traditional philosophy. For example, Leibniz enthusiastically advocated

Chinese traditional philosophy, the influences of which are branded into many parts of his philosophical system. In the west, the philosophy of Leibniz is the source for Kantian philosophy, which in turn influenced Hegel. Though not interested in Confucianism, Hegel affirmed Taoist philosophy established by Laozi and Zhuangzi. Chinese philosophy greatly influenced his dialectics, which were later developed by Marx and Engels who in turn abolished his idealist content. Therefore, Joseph Needham concluded that dialectical materialism originated from China, which was introduced by the Jesuits to Europe, theorized by Marxists and then returned to China again. In the 1960s and 70s, the celebrated physicist, Ilya Prigogine, the founder of dissipative theory, said that the founding of his thoughts benefited a lot from Chinese philosophy.

Secondly, Chinese philosophy is unique for its wisdom of perceiving men and nature, being and internal quality, knowledge of the world and self-knowledge as a dynamic dialectic process, i.e., the process of "the unity between nature and mankind". This is the exclusive thinking style and value principle of the Chinese nation as well as a legacy for the entire world, for man not only should not be in the thralldom of nature, but also should not manipulate nature unscrupulously, which is the case of the practice that "The more we dare, the more the land will yield," destroying the dynamic equipoise between nature and human beings (harmony). The entire world intensely empathizes with this view of unity, putting forward the notion of "sustainable development". As a matter of fact, traditional Chinese philosophy has embodied as well as advocated this notion since earlier times, only without those problems and scientific proofs accumulated in the process of social development. Therefore, the combination of the "double development" of Chinese philosophy and society is the only correct way, with the former presenting theorized worldview. The most basic function of dialectical materialism is to provide guidance for humans in learning and practices. As a large nation of 1.3 billion people, it is necessary that China makes more contributions to the world in the field of philosophy. One of the priorities in "combination" research is the theory of anthropological studies, which studies the unique characteristics of man and the development of his essential strength, In China, from 1949 to 1978, among the researches on Marxist philosophy, few were on human beings, while studies concentrated on men as social and class beings (that is, the common social characteristics of mankind). Actually, the quintessence of a man calls for his fulfillment as a cultured, perfected being. Thus, it is necessitous to study the development of a man and society based on the common law of individuality combined with man's collective nature (social characteristics).Marx devoted his final years to the study of this problem and it seems that it needs further in-depth elaboration. The study on man also touches upon natural sciences dealing with brain, studying the thinking and cognition of mankind. From this it can be seen that those are very vital, challenging but ineluctable questions. As to social sciences, it means to shorten the transitory historical period that runs from now to the distant future in which human beings have to work for livelihood and the fundamental purpose of work is for survival, entailing comparatively more restrictions on individuals that impede their fulfillment of personal ingenuities and talents. Certainly, work should be rewarded upon the principle of equality by society in order to exhibit fairness to individuals and care towards personal interest, thus encouraging personal aspirations and endeavors, while individuals should transcend the shackles of market exchange principles that inhibit the noble spirit of mankind. The excellence of a man should not only be equated with the anthropomorphized principles of market exchange. On one hand, limited by the level of social development, we have to accept and be "dominated" by those principles. On the other, it is obligatory for us to endeavor for the spirit that advocates "be the first to worry the woes of the people, and the last to enjoy the weal of the people," which is one of the traditional virtues of the Chinese nation and can largely shorten the rejuvenation process of China. A life dedicated to this calling can be both spiritually "liberated" and "enjoyable" in a higher level, breaking the fetters of an individual and bringing him deep-level personal fulfillment! As to social development (especially that of China), the dissemination of the spirit of patriotism and loving the general people, the important content of Chinese nation's traditional virtues, is of great significance in realities. As to philosophy, it is the important content of the dynamic development of dialectical materialism to research actual social development based on "the unity of the opposites" in combination with "the studies on human beings".

1.2.2 Relationship Between Philosophy and Natural Sciences

Though philosophy and natural sciences both study the laws of the motion of things, they are concerned with researches carried out on different levels. Every discipline of natural sciences concentrates on laws of motion in particular field or of specific objects (such as physics, chemistry, and biology). To be sure, in the process of development, different disciplines have been vertically going through layers of particularization, and meanwhile tending to develop more horizontally, embracing a broader interdisciplinary course, but each discipline is still generally one specific field of study. On the contrary, the subject of philosophy is to study the entire world, or even the most common motion laws of objects in the cosmos. To study common nature means to abandon concerns for specific contents in order to distil commonness. This abandonment indicates not considering those contents, which necessitates the conclusion that philosophy cannot replace all the specific laws as pursued by natural sciences. Therefore, the argument that "Philosophy is the science of sciences." is erroneous, but it needs to be pointed out that the two have intimate and mutually supportive bonds. Some new scientific discoveries are a testament of philosophical principles, and at the same time enrich and add to the specific contents of common laws. For example, Marx and Engels proposed time and space are the basic form of the motion of matters, but didn't specify the relations of the three. Later Professor Einstein's "theory of relativity" uncovers the interrelationship between time and space in some motions. This enriches and supplements the contents of dialectical materialism. Examples like this abound.

From another perspective, philosophy provides deep-level fundamental guidance and direction for specific disciplines. There are many cases of this. For instance, in the early 20th century, the "barber's paradox" put forward by Russell caused the third crisis for mathematics, throwing many mathematicians into confusions and even bewilderment, for paradox disturbed the strict logic system of mathematics. But if perceived from the general prospects of relative truth and absolute truth as debated in philosophy, paradox is a natural phenomenon. It illustrates our inadequate understanding of "the theory of sets" and its rules. Deepening our understanding of paradox can promote further development of mathematics. It is also concerned with the fundamental problem of mathematics whether it is a discipline that reflects the objective laws of "form" and "numbers", or merely a logic system that is derived from the logical deduction of human thinking. By the way, many disciplines of natural sciences, especially before the 1950s, have stressed upon the strict dichotomy between "affirmation" and "negation", never allowing the existence of any blurred contradictory imprecision (especially in mathematics). This attitude can be mainly attributed to the European tradition of metaphysics (which is not dialectic). Essentially, natural sciences are definitely materialist, which is undisputable, but the static metaphysical and not dialectical way of thinking have deeper repercussions. Therefore, it is desirable that more philosophical dialectical thinking methods should be injected in the further study of deep-level motion laws of various disciplines. It also has relevance to the study of brain science in the 21st century, which is the most complex, difficult, important and challenging research project because it deals with the essence of "thinking". This study naturally calls for the interdisciplinary support of philosophy and various other sciences. In the long process, crucial breakthroughs will be achieved.

1.2.3 Relationship Between System Theory and Philosophy

Both system theory and dialectical materialism admit the existence of contradictions (that is, to acknowledge the coexistence of affirmation and negation, which differs from natural sciences that don't permit contradictions and only allow unambiguous affirmations or negations in conclusions and deductions), and deem contradictions as the origin of the development of things. It is of vital importance to understand the nature of contradiction, that is, the unity of opposites. The unity of opposites is the core of the dialectic laws of the interaction of qualitative and quantitative changes, and the law of the negation of negation which play a crucial part in the existence and development of a system: qualitative changes (including mutations) are important basis for the formation of the whole by parts, the non-addition and the unique characteristics. "Mutations" include "bifurcation" and are also concerned with stability theory, cybernetics, and etc, which are significant in the qualitative analysis and precisely controlled research on laws and techniques in the study of systematic motion. The importance of the law of the negation of negation lies in the fact that the motion of "system" clearly demonstrates this law between different stages, the phenomenon of which is also exhibited in development (Things that were not feasible and could not succeed in the past can be feasible and succeed now.). The application of the law of the negation of negation in combination with actual practices is not negligible.

The concept of harmony as advocated in Chinese philosophy should be incorporated into system theory as one of the deep level objectives in the design of artificial systems. If it were only for the mechanical achievement of purposes, it would be a passive practice that lacked harmonious beauty (The optimization theory in modern operational research and cybernetics does not necessarily represent beauty.) To conclude, as a form of theorized worldview, philosophy offers guidance and basic support to the development of system theory.

One the other hand, more than a specific discipline of natural sciences, system theory enriches the content of dialectical materialism in more wide-ranging fields and on more common levels, testifying its truthfulness and thus further boosting its development. Actual cases of this kind abound and are still flourishing. Here, only several are listed for exposition. To be sure, besides system theory, cybernetics, information theories etc. all play equivalent roles. There are already some Chinese scholars who are doing research in this field (For instance, Professor Feng Guorui.)

The following is an introduction of the enrichment and supplementation provided by system theory of the knowledge, practice and theory of dialectical materialism.

(1) It enriches the content of the objects of practice as well as that of the complex multilayered practice process. When "system" is considered as an object of practice, its complexity enables the theory of knowledge and that of practice to get involved with the knowledge and practice of complex objects. This therefore enriches from the practice content to the complex multilayered practice process about complicated objects. It is an important supplement to the knowledge and practice parts of the dialectical materialist philosophy.

(2) It enriches the dialectic combination between subject and object in practices. As a science that proposes to learn the law of motion of things by systematic features, system science also provides guidance for practices through the application of those laws, which means the dialectic combination between subject and object.

(3) The incorporation of the functions of human brain into the process of knowledge and practice as a systematic process will usher in a new beginning for the dialectic combination between subject and object. From the perspective of system theory, human brain is also a system, which is only the most complicated one. By dialectically studying brain's motion process based on the a system of knowledge and practice formed more extensively, the negative effects of metaphysics and "dichotomies" will be completely eliminated. This is going to be a meaningful endeavor.

(4) The relevant contents of the development of system theory and dialectical materialism in combination can act as a bridge for the joint development of philosophy and natural sciences. Though there are collaborations in the development of philosophy and natural sciences, the two have essential differences as exhibited in their opposite directions of polarization: the former is the most common knowledge, while the latter is expertise on details in comparison, with dialectics paying most attention to the development process of motion and natural sciences stressing more on conditions and conclusions. Meanwhile, system theory possesses "amphibious characteristics". When developing in cooperation with those two "polarized" subjects, it can bridge the two.

1.2.4 Relationship Between System Theory and Other Disciplines (Natural Sciences)

The relationship between system science and other various disciplines (especially natural sciences) is mainly of mutual support and promotion, and common development, instead of that of replacement and substitution.

The general tendency of various disciplines of natural sciences is particularization, pursuing studies on detailed aspects of individualities. This is meaningful in a certain sense, but not the scene of the whole objective pattern in which things are developing and evolving towards complexity. Therefore, a fundamental problem in the theory of knowledge arises, i.e., in the long progress of mankind's constant deepening of understanding, whether the research method that stresses on particularization is enough. For instance, man is a very complex system. Then the question remains whether the pattern of studying life form is to follow the order of system, organ, cell, molecule, nucleus, atom and quark. Obviously, it is perhaps impossible to deduce the complete function of human body from the interrelationship between quarks or atoms. It further needs a holistic approach based on a certain comprehensive level. This approach is in accordance with system theory. Therefore, system theory calls for holistic research on a certain level in addition to detailed in-depth studies:

Case One: We can perceive some disciplines or group of disciplines as a system and thus study the combination of its systematic structure and secondary structures, or even carry out integrated research reversely from parts to whole. This is an important direction for the complementary development between system theory and natural sciences. Take some subjects of mathematics as an example: The combination of algebra, geometry and analysis (for instance, functional analysis), the merging of discreteness and continuity, as well as the combination between parts and whole similar to differential manifolds, are all manifestations of systematic thinking.

Case Two: System science can engage with nonlinear studies. The relations dealt with in system theory are mainly nonlinear relations, complicated by the addition of multivariate variations, so multivariate nonlinear problems pose a great difficulty, and are also the most challenging problems in the basic domain.

Case Three: We can pursue studies of various types of problem in multiple temporal and spatial dimensions (various uncertainties, such as that of probability, fuzziness, rough set theory and chaos).

Case Four: We can transform "qualitative change" into various specific types (many are nonlinear problems) to carry out extensive studies. Such as, mutation and bifurcation problems, the instability in cybernetics, interchange research in physics, chaos and fractal.

The problems dealt with by system theory all have the common nature of "complexity". There are intimate relations between the study of complex problems and system theory. Certainly, the study of complexity has wide-ranging fields with numerous problems to be researched. It is also the fundamental enigma in the development of human beings and has basic contradictions: The intelligence of mankind remains at a stable level that in millennia won't become higher, while the world is evolving in more and more complex ways, which gives rise to contradictions in man's knowledge acquirement.

Where are the possible solutions? How to represent and resolve complexity? This is both a system problem and an actual conundrum. There are numerous things to study.

To sum up, system science is nevertheless a science, and problems are naturally unavoidable in its collaboration with other sciences. It is impossible and against the laws of "development" to clearly enumerate the problems and contents. Only in the long-term process of development will solutions be found to forward its advancement, and actually solutions are development.

I personally hold that there are mainly two solutions. The first is to fully apply the law of the unity of opposites. Because a complex problem is the complicated motion of contradiction, it must be in accordance with the law of the unity of opposites. Therefore, analyzing complex problems by applying this law goes well with the principle of using basic scientific rules to solve problems. Another solution is to be found in brain science. The thinking of brain is the most complex system, the study of which will bring about the solutions to other complicated problems. Furthermore, even knowledge of part of functions of the brain thinking may further enhance the functions of brain. Though we cannot be sure about the degrees of our knowledge of workings of the brain thinking (Due to the fact that brain is involved in the study of brain, philosophical self-organization and self-evidence is touched upon.), a consensual knowledge of the brain may be achieved by the efforts of brains, study which yet needs long-term endeavors and supports from diverse disciplines. Therefore, we can deem the study of brain science as a typical example elucidating the joint development of philosophy, system science and many disciplines, which nevertheless needs further collaboration.

Chapter 2
Basics of System Theory

2.1 Definition of System and Explanations of Its Key Points

There are currently dozens of definitions for a system, on which the academic and scientific worlds haven't yet reached a consensus. Here, one kind of definition is proffered.

2.1.1 Definition of System

A system is a complex, multilayered, multi-sectional, dynamic, and comprehensive whole formed by a variety of elements, which has an open dissipative structure, constituting self-organizing mechanisms and functional relations towards the exterior.

2.1.2 Explanations of Its Key Points

1. **Open Dissipative Structure:** a nonconservative, structure not isolated that has material, energy and information exchanges with the external world, and emits negative flow of entropy toward the outside of the system.

2. **Information:** depiction and representation of the motion state of things. One representation of a motion state is called one piece of information.

3. **Self-organizing Mechanism:** a mechanism formed by the interaction between various constituents of the internal structure of a thing, as well as by a group of relations composed by the interactions between the internal structure with the external environment, evolving from unorganized chaotic state to orderly state or maintaining orderly functions. It also refers to the mechanism that maintains the orderly functions of

the system. For instance, artificial system is designed and manufactured by human beings, but has self-organizing functional mechanism.

4. **Order:** motion laws on a certain level.

5. **System:** A concise definition of a system refers to those complexes, moving things characterized by "a system", which exist objectively.

To conclude, "System" is a summarized expression of certain important motion features of objectively existent things, and the embodiment of the evolution of the world, for evolution brings it that the structure consists of complex multiple layers and numerous interrelations (including functions that form systems with external relations, as well as internal relations). It is those interrelations that exemplify multilayered general laws of motion. Moreover, with the motion of things, those laws change as well, the phenomenon of which is termed a dynamic self-organizing mechanism. This mechanism is not monopolized by living organisms; living organisms certainly possess self-organizing mechanisms to sustain life, but non-living systems also have self-organizing characteristics, which are complex if not as mysterious as those of the living organisms. In the process of knowledge acquisition and practices, the concept of "system" is gradually formulated, which in combination with realities forms system theory. It is a considerable "leap forward" for the western world that is used to analytic thinking patterns. The following sections are engaged with system theoretical research on laws of the motion of systems, which are yet far removed from in-depth knowledge of the self-organizing functions of living mechanisms.

2.2 Categories of the Unity of Opposites Often Concerned with the Application of System Theory

Systems have properties embodied by various categories of the unity of opposites, revealing complex systematic motions of multiple sections, layers, subsystems, processes, phases, and interrelations. Those several categories of the unity of opposites discussed in the following passages possess contradictions formed by incompatible extreme concepts, which however in reality exhibit various types of "unities." It is these unities of incompatibly opposite domains that have more significant bearings on the understanding of complex systematic motions. In actual practice, there are many ways to represent the unity of incompatible opposites in complex things mainly as follows:

Approximate singular representation of the dominant side of certain pair of binary opposition, even if the other side still exists, only that it is negligible on certain conditions. However, it is very important to register its existence in deep-level understanding.

In multi-sectional complex motions, each section may have different oppositional properties, so the all-inclusive understanding of the unity on the oppositional properties of different sections is imperative! The changes in the temporal domains of things entail that incompatible oppositional qualities often emerge at different time spans.

The state of a thing in motion is the temporary unity between its dominant and oppositional properties. It moves in the opposite direction. Therefore, the categories of oppositional incompatibility are indispensable in the contemplation on the unity of opposites of the motion of things.

The following discussions will elaborate on the above-mentioned contents with the help of examples.

2.2.1 Relativity and Absoluteness

This is the pair of the unity of opposites most disputed in the philosophical world. For a long time, it is argued about separately, and it is major progress made by human beings in realizing its dialectical nature. It has broad denotations and extremely universal connotations and consequently belongs to the general categories of philosophy. It has connections with the following pairs of opposites, such as "finiteness and infinity" "temporality and eternity" "whole and part" "conditionality and unconditionality" and "continuation and interruption", some of which can even be incorporated into it. Moreover, it stands for one of the most universal of the basic characteristics of the motion of objective things, with "absoluteness" representing the unconditional, the unlimited, the most inclusive of the totality and summation of things, an eternity in the temporal dimensions that never ends, independence and the strictest affirmation, etc., and "relativity" indicating all that are opposite to the above-elucidated properties. However, we should note that the unity of opposites of "absoluteness" and "relativity" has profound dialectical implications. For instance, the law that "motion represents the existence of things" is commonly applicable, unconditional, and thus absolute, but its generalness is extracted from the motions of many specific things, and consequently belongs to numerous "relativities". Furthermore, a more concrete understanding of a thing also calls for studies on its relative motion characteristics.

In mathematics, the most basic calculation "one plus one equals two" is absolute, which is the ultimate expression of abstraction. However, Egg A plus Egg B equals neither two Egg As nor two Egg Bs (on account of the fact they are not identical). They are merely two different eggs put together. When we say they are two eggs, it is only an expression of relative approximation.

2.2.2 Similarity (Resemblance) and Difference (the Comparison Between the Characteristics of Things)

The literal meanings denoted by similarity and difference are in absolute contrast, but the juxtaposition of the two points to the dialectical significances of the important unity of opposites, such as "similarity" represents "category" or the existence of groups, while only difference can designate the existence of individuals; "similarity" in certain sense means "commonness", while "difference" signifies "particularity".It is impossible to develop novel things if no particularity shows up, while if things are not incorporated into common understandings, there will be no actual advancement. The unity of opposites as exemplified by similarity and difference is also exhibited in the following ways: Things can be similar on common levels, but different on specified levels; on certain sections, they display similarities, while at the same time they manifest differences on other sections, i. e., in different "spatial" scopes "similarity" and "difference" simultaneously exist, which is also true on temporal dimensions, etc. The connotations of both "similarity" and "difference" have the implications of "absoluteness". "Completely thorough absoluteness" cannot exist in realities, e.g., a man himself cannot completely be equated with himself, because he is in constant changes. All things are beings of the unity of opposites. For instance, language is typified by similarities on general levels, while containing differences. The development of language begins with differences in certain parts and ends with their coming into common understandings and applications.

2.2.3 Certainty and Contingency (the Representation of the Compliance of the Motion of Things with Laws Mastered by Mankind)

Certainty necessarily means there exist laws of existence and their applications, while contingency implies no laws or not yet knowing the laws. When certainty and contingency are discussed, it is under certain circumstances, that is, within certain spatial and temporal scopes. Naturally, the so-called "certainty" is not changeless, which is variable in accordance with situations. However, in discussing "certainty" and "contingency", it is necessary to nail down the boundaries of time and space, for all things are in constant motion and changes. There are no absolute certainty and contingency, but the dialectical combination of the two, which indicates the complexity of the objective world (The non-existence of dominant factors or the rapid change of them is also a complex manifestation of lawlessness.) Certainty coexists and is interlocked with contingency, and the two are in a state of mutual transformation. Therefore, in the research of systems, it is crucial to discern their formation and the laws guiding their complex interactions and to hold the dialectical view of the combination of certainty and contingency. The observation of problems is necessary, for in the motion of things the correlation of certainty and contingency is an objective existence.

However, it is impossible to comprehensively grasp the laws of the motion of complex things, especially when human beings are actively involved, and the representations of many laws have the elements of uncertainty, such as the properties of chaos, chances, and fuzziness. Therefore, an overdue pursuit of certainty can be difficult to achieve.

2.2.4 Purposefulness and Natural Determinism (the Representation of the Compatibility of Man's Intentions with Objective Laws and Conditions)

The difference between human beings and animals lies in the former's conscious purposefulness, instead of instinctive reactive purposefulness, and their practical abilities in the fulfillment of this purposefulness. Purposefulness implies the future, the realization of which depends on "natural determinism". The so-called "natural determinism" refers to specified restrictive limits of objective existence, beyond which men's purposes cannot be achieved. The birth of a new artificial system emblemizes certain purposes and creativities of mankind. The resolution of the relation of the unity of opposites between purposefulness and natural determinism means skillfully "agreeing" with natural determinism to the uttermost degree, which includes the transformation of practicing methods while maintaining the "purpose" as the prior condition, thus changing, and breaking through the constraints of "natural determinism". This is also one of the research contents of system theory and artificial system design.

2.2.5 Whole and Parts (the Relationship Between Holistic and Partial Existence in Things)

A very important portion of system theory concerns the study of the relationship between whole and parts. In the design of an artificial system, its overall function is the priority of consideration, but it doesn't equate with the agglomeration of the functions of various parts. The overall function which surpasses the sum of the functions of different parts added together is the non-linear transcending effect we strive for. Certainly, the opposite should be strenuously avoided. However, restricted by "natural constraints", its negative relations generally objectively exist. Thus, the content of "the whole" includes both "positive" and "negative" parts, and it is important to correctly appraise and avoid the transformation to negative effects. The relationship between parts and whole is also comparative. The whole which covers certain scopes can normally be incorporated as a part of more extensive scopes. In addition, parts can also change into the whole. The evolution, decline, and fall of things on many occasions are initiated by parts that then lead to overall changes. Therefore, the unity of opposites is embodied by whole, and parts is one of the spotlighted attractions that interest us.

2.2.6 Complexity and Simplicity (General Characteristics of Mankind's Knowledge and Assessment of the Motion and Existence of Things)

This pair of binary oppositions is typified by the interfusing and conditionally mutual transformations of "absoluteness" and "relativity". Complexity is the basic feature that characterizes the existence and evolution of the world or even the entire cosmos. It is also the epitome of absoluteness, which however contains many contents that form the unity of opposites with "simplicity", for example, the formulas of commonly applicable laws are simple, but their actual usages are complex; or certain things are complex to some people, but simple to those who are familiarized with these things. Because of the progress of human beings, things that used to be complex in the past are now simple, etc. Men are themselves complex entities, who nevertheless often evade complexity in search of simplicity to smoothly solve problems, but on some occasions, they endeavor to understand and grasp complexity. "Simplicity" is the opposite of complexity, which nonetheless is as variable as complexity. In system theory, what is to emphasize and take note of is the multifaceted dialectical "transformations" between complexity and simplicity. The objects of the practices of "systems" are complex, and the important laws of their motions are targeted by system theory, which is to be distilled and condensed from large amounts of complex phenomena. Underlying their formulations are the rules of transformation from "complexity" to "simplicity" while applying laws to solve complex actual problems accords with the rules of transformation from "simplicity" to "complexity".

2.2.7 Qualitative and Quantitative Changes (Representing a Pair of the Most Basic Motion Characteristics in the Motion of Things)

Here only two points will be emphasized. The first is that in the survival and development of a system, there are many multilayered and multi-sectional circular processes of quantitative to qualitative changes as well as the states that interweave both (Some section is in qualitative change, while other sections can be undergoing quantitative change.) The second point is to grasp the conditions and laws for quantitative and qualitative changes (especially conditions for the latter) and utilize those conditions according to the laws to facilitate the achievement of man's goals. It embodies the progress of human beings, their creativity, and their aspirations.

2.2.8 Continuous and Discontinuous (the Representation of the Continuation of the Changes, the Adjacency of Certain Important Characteristics in the Motion of Things in Temporal and Spatial Order)

They have multiple meanings: Continuity can refer to the objective existence of the whole, while discontinuous can point to the existence of individuals; the former can imply quantitative changes, while the latter can stand for the breaking-up of quantitative changes and the emergence of qualitative changes. The dialectical transformations between continuity and discontinuity constitute the basic laws of "motion", the correct application of which exhibits not only the wisdom and capabilities of human beings but also the proper management and treatment of commonness and particularity. In the analysis and practice of "system", the concepts of "continuity" and "discontinuity" should be appropriately applied for the guidance of practices, for the motion of things comprises of the combination of "continuity" and "discontinuity", and there is no isolated "discontinuity" and "continuity". Besides the alternation of "continuity" and "discontinuity" in the order of a process, there are also motion processes that interweave spatially (in a general sense) and temporal continuity and discontinuity. For instance, directed by the mentality of "conquering nature", to increase grain production, large-scale reclamation of land is carried out to enlarge farmlands. Though grain production is gradually increased, the biological environment is destroyed, which leads to a series of meteorological deteriorations that devastate the sustenance of water and soil. Consequently, grain production cannot be augmented. Therefore, it is of great necessity to properly handle the relationship between "continuation" and "interruption."

The absolute mentality of "fighting" against nature will result in extreme measures of "dominance", which only lead to the opposite effects. In certain spatial and temporal scopes, it is somewhat meaningful to advocate the concept of "harmony" as promoted in Chinese philosophy. The pursuit of a harmonious state is necessary.

The above passages are concerned with some concepts of the unity of opposites, with each pair stressing certain aspects (section) to express the relation of the unity of opposites, but all of these "sections" are not isolated but interrelated. Therefore, these categories of the unity of opposites are interwoven and interlocked.

2.3 Unavoidable Brief Discussions on "Beauty"

2.3.1 The Essence of Men and Beauty

The essence of men lies in the fact that they can intentionally carry out purposeful and organized activities (including creations) to improve their living environment, while that of beauty (not equivalent to that of men) consists of the affirmation, verification, and acclamation of the essence of men, or more specifically, the objectified praises of the essential forces of men, that is, the embodiment of men's creativity, courageousness and intelligence in the demonstration of their essential powers in certain social relations, as well as the validation and compliments of the bounties of civilization (spiritual and material) accumulated through men's social practices. Once in the above-illustrated circumstances, men will obtain a genuine feeling of passion, pride, and pleasure because of the perception of their own essential powers. This is the so-called sense of beauty, which is a commonly possessed complex emotion of mankind that varies from individuals, who nevertheless instinctively love and aspire to it. To be more precise, beauty can arise from both direct and deep-level perception of objects (such as the perception of hidden contrasts, high-level learning powers, etc.)

2.3.2 The Categorization of Beauty

Beauty can be approximately categorized into formal beauty, artistic beauty, social beauty, natural beauty, scientific beauty, technical beauty, engineering beauty, etc. Each category can be further divided and when interwoven with each other come diversified integrated beauties.

For instance, formal beauty can further be divided into the beauty of the organic whole, symmetry, coordination, diversification, proportion, and rhythm. It is attached to the immediate object, so it is both direct and indirect (indirect as to form).

Or take natural beauty for example. It encompasses natural formal beauty as well as indirect perceptive beauty, which refers to the commendation of mankind's natural perceptive abilities (not the objects directly created by man).

Neither scientific nor technical beauty stimulates man's sense of happiness simply through actual entities. Instead, they engender an indirect sense of beauty based on mankind's understanding of the importance and difficulty of scientific and technical problems as well as the ingenuity evinced in their solutions. The following short passage will be devoted to elaborate discussions of scientific, technical, and engineering beauty.

2.3.3 Scientific Beauty

Scientific beauty is a kind of complex beauty, which arises from the acknowledgment and interaction of truth, goodness, and hardship. To sum up, it is a kind of complex and comprehensive beauty.

Normally, the broader its scope, the more common its applicability, a scientific law will be more difficult to discover and confirm. Besides, there are also laws of scientific issues which are equally difficult to discover and confirm because of their complexity. In addition, scientific laws endow mankind with more freedom, boosting their development and evolution. Therefore, they are mainly "good" on this account. Major scientific discoveries call for dexterous thinking and methodology.

Each discipline has its own unique content, thus constituting wide-ranging content of beauty. Moreover, different disciplines form intricate intersections, whose resolution often embody "beauty" which can include complex contents and simple representational forms (which is generally the case of scientific laws), with "simple" forms augmenting the sense of beauty. To conclude, the perception of scientific beauty is based on mankind's advanced acuity, which is a unique scientific enjoyment, only to be gleaned and tasted by those scientists who have after arduous endeavors finally achieved success. By the way, for juveniles to learn the rudimentary of scientific beauty, aesthetic education, scientific education, and outstanding scientific works are indispensable. The following are two case studies for the explication of scientific beauty:

Case 1: the profile analysis of the beauty of the unity of opposites

First, there is the beauty of the simple form of its expression, which has only five Chinese characteristics, with "lv" or "law" representing law and truthfulness, and the other two nouns formed by two characters illustrating the unity of opposites in the motion of things. It is full of exciting meanings and profundity, but much uncomplicated, which is therefore admirable, astounding, and exhilarating. This is called the sense of beauty, which is a kind of beauty. To add to this, there is its meaning of profound significance, and two-charactered words of equal importance, constituting multilayered implications and enhancing its general depth as a law. For example, the meaning of "opposites" consists of diverse contents, such as contradiction, mutual negation and confrontation, etc., "unity" denotes mutual reliance, transformation and concordance on a more general plane, etc., while "the law of the unity of opposites" holistically reveals the essence of the motion and development of things in the simplest form and within the most common scopes, the profundity and dialectical completeness (including the infinitely expansible nature of its own content) of which make people acutely experience its beauty and excitement in the appreciation and application of it, and genuinely pour their compliments on it. This is called deep-level beauty.

Case 2: The Special Theory of Relativity. It has resolved the deep-level contradiction between the theory of electromagnetic fields based on Maxwell's equations and Newton's mechanics. This contradiction is concerned with the nature of the system and existence (through the invariance of the velocity of light). The concepts of relative time, space, and their relatedness with velocity as postulated in the Theory of Relativity complement classic dialectical materialism which didn't touch upon a time, space, and their relations with motion, and also correct the errors committed by mankind who have validated the absoluteness of time and space based on direct simple experiences.

To sum up, the depth, universality and significance of scientific knowledge are correlated with the simplicity of its formulation, causing us to perceive its multilayered and profound beauty!

2.3.4 Technical Beauty

Technical beauty is also a kind of deep-level, indirectly sensed beauty, which includes many aesthetic factors. Owing to the diversity of categories and contents of technology, the content of technical beauty is also very rich and plenteous. The basis of technology originates from science and the adroitness of mankind's thinking. Here the essential elements of technical beauty, including the beauty embodied in the significance and importance of the solutions of the problems encountered in the practices of human beings, the resourcefulness of those solutions, and the connection and harmony between science and technology, etc., are only sketchily dealt with.

2.3.5 Engineering Beauty

"Engineering" has two meanings, one referring to the disciplines of engineering, such as electronic engineering, mechanic engineering, etc., the other referring to certain specific engineering projects in construction, which are mainly large-scale projects of great significance, built up to serve people. With the advancement of human civilization, various complex immense projects of momentousness have been burgeoning, the failures of which will often entail enormous losses or even severe catastrophes. Therefore, studies on how to succeed in complicated major projects have drawn increasing attention. Diverse comprehensive disciplines are involved in engineering, encompassing technical, scientific, organizational and implementation subjects, as well as interdisciplinary ones combining science and technology, etc. Engineering disciplines usually incorporate science, technology and the beauty embodied in their combination. Some disciplines, e.g., architecture, stress particularly on formal beauty, while some interdisciplinary subjects coalescing engineering and art, for instance, design, naturally put priority on aesthetics and conjoin this with the objects of design. As to specific engineering projects, there are many factors which arouse people's sense of beauty, such as the formal beauty demonstrated by the projects (magnificence, harmony…), the mixture of natural beauty with formal beauty exhibited in the combination with natural environment. In addition, the important functions provided by those projects harmoniously put various techniques into application, and ingeniously solve crucial or difficult problems encountered in practices, evoking feelings of praise, adoration, excitement, acclamation, pleasure, comfort from scientific workers and those who know the truth, or have enjoyed the service provided by engineering projects. All of these are beauty, which is comprehensive, deep-level and indirect. This kind of beauty perceived by people in the harmonious combination of the functions and structures of engineering projects is the main content of engineering beauty.

2.3.6 Scientific and Engineering Beauty in the Fields of System

The scientific beauty of system theory covers many contents. In completion (now it is far from completion), this discipline will possess various kinds of beauty owned by the law of the unity of opposites, as well as the true, the good and the beautiful. Besides, it will have the beauty of the crafty combination of scientific laws and aesthetics. At present, it is not easy to clarify, but in the future, when the systematic structure of system theory is finalized (which, however, is not the end of this discipline), scientists will be awakened to its beauty.

Engineering systems are normally more complicated and important projects. As a result, they have engineering beauty as well as many contents of technical beauty. The creation of a significant, technically advanced, complex and large-scale engineering system necessitates scientific supports and specialized techniques of diverse disciplines. This kind of sciences are not "pure" sciences (which target the discovery of scientific laws), but sciences that are combined with technology, which therefore are now termed technical sciences (These are disciplines of great importance.) They have the charms of science and technology in combination, which naturally correspond with certain kind of beauty. From the perspective of civilization and advancement, we yearn for more of this kind of beauty, which creates in us the sense of beauty, in order not to be constrained by the necessities of everyday life.

2.4 Axioms of System

In a relatively important discipline, some basic concepts are put forward, based on which relevant thinking methods targeting problems are raised for their analysis and judgment in order to logically deduce some new laws and conclusions; after many rounds of verification and development, a system of scientific laws in a certain field is formed (Axioms included). This is the formation of a discipline (which undergoes many daunting adversities in the process). The contents illustrated above have discussed the academic domains, crucial problems and related key concepts, in an attempt to establish a system of axioms, which must be free of self-contradictions, independent, all-inclusive and not plagued by redundancies. No self-contradiction means that by applying different articles of this system of axioms self-contradictory conclusions shouldn't be deduced. "Independent" refers to that articles within the system shouldn't be deduced from other articles of the system, while all-inclusiveness means that all the conclusive contents of the discipline can be deduced from this system of axioms. Finally, "free of redundancy" implies that there are no redundant articles in the system. Though the establishment of the system of axioms needs no verification, its accuracy and comprehensiveness need constant testament and perfection in a very long process, especially when the burgeoning of new disciplines and the mutual influences between different disciplines lead to problems of compatibility and comprehensiveness on deeper levels and broader scopes. For instance, the combination of Newton mechanics (including Galilean transformation of velocity) and electromagnetics (Maxwell equations are complete, while the speed of light commonly appears as C.) engenders the problem of compatibility: either some amendments and supplements are introduced in Newton mechanics, or electromagnetics is restricted to the inertia system (But the discipline of electromagnetics is complete and universally applicable, which needs not to be confirmed in inertia system.) This problem had existed and been disputed for years before Special Relativity Theory put forward that Newton mechanics is only applicable in low-speed circumstances, resolving the problem of compatibility and demonstrating that human beings are gradually learning truth.

2.4.1 Top Axioms of System Theory ("Shared" with Materialist Dialectics)

Because system theory stresses on the dynamic motions of things on integral levels, it has many relations with philosophy, especially materialist dialectics. The basic laws of materialism can be considered as the top axioms of system theory, i.e., the laws of the unity of opposites, of quantitative and qualitative changes, of the negation of negation, among which the law of the unity of opposites is more commonly applicable.

2.4.2 Temporary System of Axioms of System Theory

2.4.2.1 Temporary Axioms of System

Any system must be embedded in a more extensive system as its subsystem, and itself contains several subsystems (which can be intertwined), so the composition of the system for the multi-level interwoven structural system (here, the more extensive system is not limited to the geometric space of the larger, but the broader sense of the more complex functional structure, more universal influence relationships, such as the evolution of movement development of the larger sense). In the field of systems theory and artificial systems design, we believe that the axiomatic system of systems theory consists mainly of the following axioms:

Axiom 1: Living systems are all non-conservative dissipative systems, which have material, energy, and information exchanges with the external environment.

Axiom 2: The fundamental reason for the survival of system lies in the motion of the unity of opposites formed by diverse interrelations of the "functions", "structures" of the system and environment (with other systems).

Axiom 3: A living system must be dynamic and orderly on macroscopic levels. "Order" represents overall characteristics of the motion of system. On some occasions it can be further condensed to be signified by order parameter. It is the embodiment of the dynamic self-organizing mechanism formed by the interactions between system "structure" and external environment, which in the process of motion undergoes constant quantitative and qualitative changes and reaches new order through fluctuations.

Axiom 4: A system and its environment evolve together, the evolving process of which also incessantly evolves, from low level, comparatively simplicity, to high level and complexity. It is a heterogeneous and nonlinear complex motion process in time and space, in compliance with the law of the survival of the fittest.

Axiom 5: The motion of systems (mainly exhibited in functions and structures) is a kind of material motion. Therefore, there must be conditions and constraints, i.e., those of human, physical and biological reasons as well as the realities of things. Because system is law-abiding material motions in the open environment, there must be both "acquisitions" and "expenses" for the application of system and itself.

2.4.2.2 Brief Explanations of Axioms of the System

About Axiom 1, it is concerned with whether a system can be infinitely divided, which is a physical as well as a philosophical problem and is still disputed at present. Take quark, for instance: Some physicians argue that it can be infinitely divided, while others consider it closed. Here, in physics, we are only concerned with subdivisions under the state of interwoven functions and structures instead of subdivisions such as molecules, atoms, nuclei, etc, in physics.

About the specified explanations of the "internal causes" of the existence and development of a system as a thing, there are few detailed analyses and explications in philosophy. In the system of axioms of system theory, what are pointed out are the internal structures as well as the organization of the related parts of existent internal and external relation systems (including functions).

The system of axioms stressfully puts forward that "system" is a dynamic moving thing, which is characterized by motion laws of dynamic processes. It means that the interactions of relevant "relations" form regular motions and when the former change to a certain degree, indicating changes of rules, the order of the system will change accordingly, which is qualitative change. Once the "conditions" for qualitative change are ripe, a random perturbation will precipitate this change.

In the past, it was believed that evolution was to be understood in accordance with "the theory of selection", that is, things were passively selected by the environment to be eliminated or evolve. Now systems and the environment are perceived to be evolving together in interactions (systems also influence the environment). Moreover, the mode and mechanism of evolution are also developing and evolving, which is termed as the evolution of evolving processes.

"Systematic existence" is a kind of material motion, which calls for certain conditions. By making use of certain rules, mankind can change the state of motion of "systems" (in general sense), but have to "pay the prices" in the process. The change of the state of "systems" according to the purposes of human beings is called "acquisition" ("acquisition" in general sense), and the prices they pay are called "expenses". This is a universal law.

2.5 The Expanded Model of Multilayered Temporal and Spatial Domains of Systematic Motion

The functions of the establishment of an expanded model: the multilayered expansion and representation of the motion process formed in the core mechanism of system in the temporal domain and the general spatial domain, based on which the research (analysis and synthesis) on system can be conveniently carried out.

2.5.1 A Basic Theoretical Framework Behind the Establishment of This Model is Composed of the Following Core "Elements"

➢ Basic theories of modern system theory:

Such as the dissipative structure—self-organization theory, synergetics, mutation theories of Open Complex Giant System, as well as more general laws of the unity of opposites, and axiom systems temporarily established.

➢ Concepts Representing the Essence of System Motion

"Relationship", "dissipation", "self-organizing features", temporal and spatial multilayered open interweaving functions and structural features, "macro-sequence", function, structure, "constraints", various domains of the unity of opposites (such as absoluteness and relativity, quantitative and qualitative changes, continuation and interruption, purposefulness and natural determinism, commonness, and particularity, etc.).

➢ Applicable Methods in System:

The combination of qualitative and quantitative methods, meta-relation group methods into which specific "relations" can be flexibly incorporated.

➢ Division of the sectional expansion of time and space:

Space: system (including sections), secondary system (section), subsystem (section) ... element (section)

Time: period (which is divided into incubation, birth, development, maturity, decline, rejuvenation, aging), process, phase, sub-phase, fragments, minute fragments.

2.5.2 Multilayered Multi-periodical Expansion Model of System Based on Relations (Represented Separately in Terms of Space and Time)

This section first gives a schematic multi-level spatial domain unfolding and a time-domain unfolding model of system motion based on relationships, respectively, and finally deals with the time-domain unfolding of system survival and the schematic model of system spatial structure embedding.

2.5.2.1 Structure Expansion Diagram ("Relation" as the Links)

Based on the above discussion, a schematic representation of the system's hierarchical unfolding of the spatial structure linked by relationships is shown in Figure 2-1.

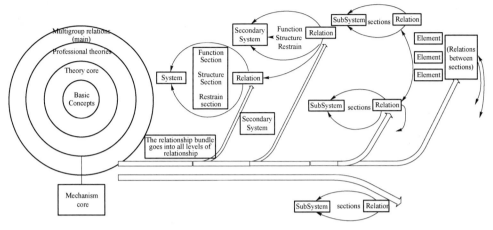

Figure 2-1 Schematic representation of the hierarchical unfolding of the system's spatial structure linked by relationships

2.5.2.2 Topological Structure Diagram of Sectional Relations of System Spatial Structure

"Relations" can be not only hard structural material relations, but also information exchanges, motion interaction effects and other soft relations, as well as the combination of soft and hard relations. See Figure 2-2.

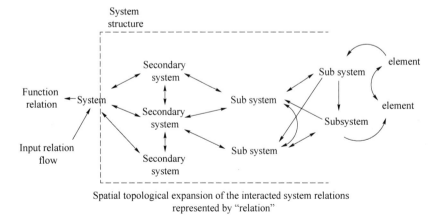

Figure 2-2 Topology of the relational structure of systems characterising interactions in terms of relationships

2.5.2.3 Embedded Diagram of System Survival Expansion in Temporal Domain and System Spatial Structure

The time-domain unfolding of the system survival and the system space embedding are shown schematically in Figure 2-3.

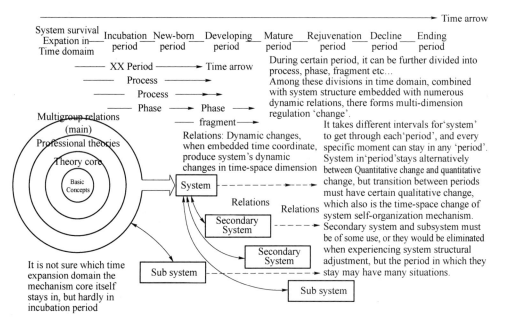

Figure 2-3 Schematic representation of the time-domain unfolding and system space embedding for system survival

In the incubation period, only related theoretical system and basic system framework are embedded into the systematic survival motion process. In the newborn period, the multi-level spatial structure of the system in the developmental period is first initially then basically formed, which is embedded into the systematic survival motion process. See Figure 2-4. In maturity, the complete spatial structure of system as demonstrated in Figure 2-5 is embedded into the system motion survival process, allowing the system to maintain orderly functions. In the declining and rejuvenating stage, system spatial structure will undergo adjustments, removing parts that should be eliminated and adding structures of self-organizing mechanism in order to meet the requirements of rejuvenation.

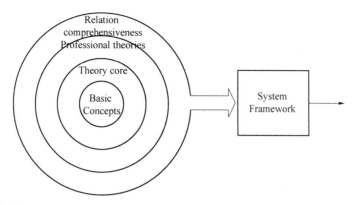

Figure 2-4 Multi-level spatial structure of the system in its infancy and development

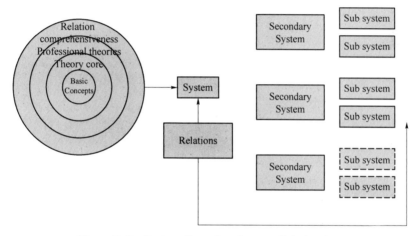

Figure 2-5 Systematic and mature spatial structure

The survival process of a complex system consists of the interactions of the elements, components, sub-subsystems, subsystems, and secondary systems of the system and itself, which expand on the temporal dimension as organized and regular motions represented by relation groups. The laws of this kind of motion are termed "order". Because all parts of the system are in regular motion, a variety of multilayered dynamic "orders" exist in the system. Those on the overall level of the system are referred to as the macro "orders", which are organically composed of "orders" of different parts. System theory is to study the conditions and rules under which the system generates motions, while artificial system design comprises extended studies and research on new systems (artificial systems) with a higher success rate based on system theory (which will be featured in later discussions).

2.6 Relations

"Relation": an important and core concept in system theory, which as the breakthrough point in qualitative and quantitative research is applied to more specific studies on the motion of "system".

2.6.1 Mathematic Definitions

Suppose A is a set, D is a binary set [right, wrong], a A × A to D mapping R is defined as a relation between the elements of A, then if a, b ∈ A, aRb →right, then a, b accord with Relation R (aRb); if aRb →wrong, then a, b do not accord with Relation R.

∴ "Relation" is mapping.

Equivalence relation in mathematics: represented by the symbol $\backsim$, with three characteristics in total, namely:

a $\backsim$ a, reflexivity;

a $\backsim$ b then b $\backsim$ a, symmetry;

a $\backsim$ b, b $\backsim$ c then a $\backsim$ c, transitivity;

Note that c is Γa, Γb, or a $\backsim$ a will lose independence.

The significance of equivalence relation lies in its role as the criterion for the division of set, which has the following two theorems:

"The subset divided from set A must have a corresponding equivalence relation"-mathematical theorem in natural language. In order to prove this mathematical theorem, mathematical method must be applied: it has first to be turned to mathematical language and logical framework, and then to be proved by existing mathematical knowledge (theorems, etc.)

Theorem: In a set, the division of a subset corresponds with an equivalence relation

Suppose the division of a subset is in accordance with a criterion, those that meet the criterion are allowed into the subset, and this criterion is $\backsim$

Elements in the subset a $\backsim$ a, b $\backsim$ b, c $\backsim$ c

a, b, c all accord with $\backsim$, ∴ a $\backsim$ b, b $\backsim$ c, a $\backsim$ c

b $\backsim$ a, c $\backsim$ b, c $\backsim$ a, $\backsim$ has symmetry, transitivity, reflexivity, therefore $\backsim$ is equivalence relation.

Theorem: an equivalence relation can be divided into a subset

Suppose A is a set, equivalence relation is $\backsim$, based on reflexivity, a that accords with a $\backsim$ a can be arbitrarily singled out from A. From a according to a $\backsim$ b... pick b, c... a subset is formed.

As a $\backsim$ b, b $\backsim$ a, ∴ it makes no difference whether picking a, or b first, which can be extended to the fact that arbitrary selection makes no difference

As a $\backsim$ a, b $\backsim$ b, c $\backsim$ c, a $\backsim$ b, b $\backsim$ c, therefore a $\backsim$ c, demonstrating that applying $\backsim$ to select elements can divide subsets, isomorphic relations, homomorphic relations (such as a $\backsim$ b, b $\backsim$ c, c $\backsim$ d... serial chain).

∴ the completeness of the theorem can be proved by steps 2 and 3, i.e., subsets can be divided by using equivalence relations, with no need of other conditions.

Other mathematical theorems are not applied in the verification of the above two theorems, the method of which is the simplest.

Similarity relation in mathematics: transitivity is detracted in equivalence; it is a relation whose constraints are weaker than those of the equivalence relation; "general similarity" cannot have transitivity. A son looks like his father as well his mother, but it doesn't entail that there is resemblance between the two parents.

Isomorphic and homomorphic relation:

Suppose Set A and Set A' respectively consist of elements $a_1 a_2 \ldots a_n$; $a'_1 a'_2 \ldots a'_n$, each respectively has operations o and $\bar{o}$, and $A \rightarrow A'$, mapping is ϕ, if:

$\phi(a_1 o a_n) = \phi(a_1) \bar{o} \phi(a_n) = a'_1 \bar{o} a'_n$, then ϕ is called homomorphic mapping (as to A and A' in algebraic operation o and $\bar{o}$, A and A' are homomorphic).

If ϕ is added as surjection, then ϕ is called surjective homomorphism;

If ϕ is a mapping, then ϕ is isomorphic (as to A and A' in algebraic operation o and $\bar{o}$, A and A' are isomorphic);

If A and A' are isomorphic, it means that operation o between the elements in A corresponds with operation $\bar{o}$ between the elements in A', which has the same nature, but only different forms. Isomorphic mappings have many uses in the analysis of system.

2.6.2 Connotations of Relation in System Theory

The connotations of "relation" defined and applied in system theory are comparatively broad, embodying the far-reaching complex connections between things.

The various interactions between things are termed relations;

The existential states (which are also called relational states) in consequence to the various interactions between things;

The temporal and spatial comparative states between things;

The transfer of certain existential state of things under the stimulation of certain forces.

2.6.3 Some Important Relations Concerned with in System Theory

✦Functional Relation

It refers to the main interactions between system and "environment", which basically form those relations that provide specific services for some external objects. Those relations are termed functions, which make possible the survival of systems. They are actually constituted of many multilayered "relations", the dynamic sum total of the functions (also relations) divided by subsystems and secondary systems.

✦Structural (relation)

The inner constituents of a system make up its structures, which as a matter of fact besides the hardware of secondary systems and subsystems have interrelations between as well within themselves (some are software). Some of them are the "functions" of subsystems and secondary systems. It is these interrelations (functions) that form the dynamic existence of systematic structures, which are meanwhile the "functional" basis on the level of system as well. We can assume that the "structures" of a system and the important internal and external relations are the basis of the existence of system, i.e., the internal causes of things in philosophical terms.

✦Constraint Relation

In general, the realization and existence of "functions" and "structures" are conditional. The prerequisite and restrictive conditions of their realization are called constraint relations. These relations originate from human, physical and biological reasons as well as the realities of things, the strictest and most insurmountable of which are the limits set by the legal system. Constraint relations can also appear in the form of rules or the states and levels of technology. Their restrictive effects are dissolved in the spatial and temporal motions between objects, and take diverse forms (For instance, they can be not only continuous effects on temporal dimension, but also discontinuous ones on different phases, e.g., part of the constraints of law are punishments for producing law-breaking results.) Moreover, it should be pointed out that constraint relations have systematic features, such as, the constraints of "constraints", different levels, sections, times or restrictive conditions. To conclude, constraint relation is a kind of relation, which exerts constraining effects on a certain function in the interactions of things.

2.6.4 The Classification of "Relations" from the Perspective of the Connotations and Denotations of "the Concept of Relation"

Subordinate Relation: If the denotations of one "relation" are all covered by and become a part of those of another's, then the one with broader denotations is called generic relation, and the other auxiliary relation, but their connotations are in direct contrast, with the former having narrower connotations that are covered by the latter's, which in addition has connotations of its own.

Parallel Relation: in a generic relation, referring to those auxiliary relations that respectively have exclusive denotations (which don't overlap).

Cross Relation: "relations" that have connotations and denotations which partly overlap.

Identical Relation: two relations that have the same denotations but different or not completely different connotations (embodying the multi-sectional nature of things)

Contrasting Relation: a special form, that is, in parallel relations, auxiliary relations at two opposites that have contrasting denotations.

Contradictory Relation: a special kind of parallel and contrasting relation, with the total of the denotations of the two auxiliary relations equating those of the generic relations.

Functional relations, structural relations, and constraint relations in combination with specific systems (specific systematic structures) can have the above relations.

2.6.5 Relations of the Objects Sustained by "Relations" (Set Objects as XYZ)

Similar Relation: X is similar to Y.

Temporal and Spatial Comparative Relation: X is bigger than Y, X is preceding Y, X is earlier than Y, later, ..., as well as temporal and spatial combinations.

Possessive Relation: X has Y, X controls Y.

"Be" Relation: X is Y.

Hereditary Relation: certain characteristic of X is passed down to Y, Y inherits the characteristic of X.

Cause-effect Relation: X is cause, Y is effect, Y is engendered by X.

Contradictory Relation: X and Y has contradicting relation.

Member relation, partial relation, component relation, group relation, etc.

2.6.6 Spatial and Temporal Expansion of Relations

Whatever form the motion takes, it must be expanded in time and space, while space refers to general space, which is not restricted to three-dimensional geometric space. Moreover, it is not only expanded in time or space, but also dynamically expanded in the combination of temporal and spatial dimensions. Dynamic expansion can be represented by multiple temporal and spatial fragments, as well as the rate of changes in relations.

In continuous mode, $\frac{\partial}{\partial t}, \frac{\partial^2}{\partial t^2}, \frac{\partial}{\partial x}, \frac{\partial^2}{\partial x^2}, \nabla, \nabla^2$ can be combined to constitute a dynamic expressive system, while in discontinuous mode, a sequence can be applied.

If a certain section of the relation, is equivalent, this section can be perceived as similar, though other sections can be not equivalent. It is this characteristic that can be used for transformations and transfers. For instance, relations of equivalent "functions" and not equivalent "constraints" are typical of those. This kind of "transformations" can be seen as a dynamic transfer in temporal and spatial expansion.

In addition, spatial contrasting or contradictory relations can in the temporal dimension make contradictory components exist not simultaneously in accumulations, forming transfers of contradiction, which is also a dynamic transfer in temporal and spatial expansion.

For example, flood volume has contradictory relations with the net capacity of several reservoirs, but the downstream reservoirs can adjust capacity through dynamic drainage (under the premise of maximum flow allowed by the river channel), in order that conflicts of capacity don't occur with the flood simultaneously. This is a wise "temporal and spatial expansion".

2.7 Ideal Model of System and Its Application

See Figure 2-6. The most scientific thinking pattern of mankind is that of dialectical materialism, the application of which in the process of resolving difficult problems also calls for mankind's concrete, detailed, qualitative, and quantitative thinking in a multi-

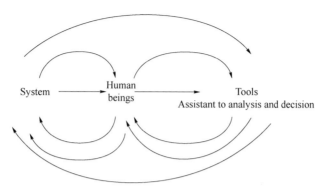

Figure 2-6   Dynamic Circular Circle of System-Human-Auxiliary Tools

dimensional combination with realities within time and space. However, confronted with the sheer colossal number of questions, mankind's abilities are far from adequate. Only through tools, especially through the help of information processing instruments, the most effective of which is the computer, can this deficiency be redressed. However, the foremost shortcoming computer is plagued with is its lack of thinking functions. It can only implement those orders after men have formularized problems and designed solutions (computing methods), under which circumstances the computer merely plays the role of a minor and insignificant tool. Equipping computers with some thinking abilities can transform them into mankind's advanced tools, though the realization of this work is tremendously arduous, which cannot be achieved in three to five years. It may last for the length of centuries (It demands the support of enormously complicated, now unrealizable scientific and technological systems, such as cognition science, brain science, nonlinear theory, and finite management of massive objects…), but the harmonious combination of human-computer-systematic objects is the inexorable ideal of development. Currently, there are two routes for the development of the combination of systematic objects-human-computer (research and applications by combining qualitative and quantitative methods): one is revolutionary and fundamental, that is, to penetrate the thinking mechanism of the human brain, based on which a computer system with certain thinking abilities is constructed, making the combination of system-human-computer more or less ideally; the other refers to step-by-step improvements, which though not thoroughgoing can prove to be very fruitful and meanwhile presents lots of work to be dealt with. This second route is also very difficult. It is a kind of methodological system, i.e., modeling simulated computing and analyzing platform (where a man plays the dominant role), which is at present finding increasingly extensive applications.

2.8 Thoughts on the "Improvement" of the Efficiency of Computers

It is a path worth our strenuous efforts to improve the performance of computers as tools through the establishment of a formal language expressive framework based on the law of the unity of opposites, in further combination with high-order predicate logic, modal logic or fuzzy logic operations.

The expressive framework of the unity of opposites demonstrates:

The main characteristics of a thing are decided by the major aspects of its internal contradictions.

Contradictions of a thing comprise opposing sides, with one dominating the other.

The existence of contradictions is based on the prerequisite (A1)

Opposition is as follows: mutual counteracting, weakening, and negating.

Unity is as follows: identical and interdependent in broader denotations

Motion laws of both sides of the contradiction: (occurring under the presumption of the occurrence of A2):

Conversion of major aspects.

Common development as a joint entity.

To mutually use parts of the factors of each other for self-development.

Main models of the impacts of the law of the unity of opposites on the development of things:

One eliminates the other, transforming or further deepening the nature of things.

Both perish, and new things are born.

The two fuse for the formation of new things or new stages.

Note: the establishment of atomic propositions or names by referring to "permission".

Chapter 3
The Current Development of System Theory

3.1 Dissipative Structure—Self-Organizing System Theory and Its Systematic Framework

Professor Prigogine's dissipative structure theory is the theoretical basis for the dissipative structure—self-organizing system theory, with the significance of the first principle, due to the fact that the dissipative structure forms the basic condition (the necessary condition) for the construction of the self-organizing features of a system, which are the most basic characteristics of a living system. Only after these two important basic features of the system are studied can more scientifically research on the evolution of the system (the core mechanism of evolution-the generation of a new "order" and the formation of a new evolutionary process) be viable. From the opposing interactions between evolution and the "environment", the dissipative structure—self-organization theory can be further extended to the study of more complex inter-related systems. On this account, the dissipative structure—self-organization theory is considered one of the foundations for the development of modern system theories. What is more, the dissipative structure—self-organization theory has also been developed into a system, which in the author's opinion is composed of the dissipative structure—self-organization theory of Professor Prigogine, the synergetic of Professor Haken, the self-organizing cosmetology theory of Professor Jantsch and the concept of Open Complex Giant System of Professor Qian Xuesen, in addition with stability theory, bifurcation theory, positive feedback theory and other existent theories concerning researches on the changes of self-organizing functions and "order" (order parameters) of a system due to environmental changes and "landing" effects. To the limitation of space in this book, association theory is not discussed.

3.2 Dissipative Structure–Self-Organization Theory

3.2.1 Dissipative Theory Proposed by Professor Prigogine and the Theoretical Basics of the Generation of Minimum Entropy

According to their relationship with the "environment", the systems can theoretically be divided into conservative systems that have no relationship with the environment (i.e., these that have no matter, energy, and information exchanges with the "environment") and open systems. After a lapse of time, the conservative systems reach a state of equilibrium. It is at this stage that a conservative system arrives at the maximum entropy according to the second law of thermodynamics (without entropy flow), while its state functions are distributed evenly in space, which is invariable in time, that is, constant in both space and time. Professor Prigogine pointed out that the macro-order of a conservative system is zero, which means macroscopic disorder. Therefore, we believe that conservative systems are meaningless in the investigation of the evolution of systems, which need no further elaboration.

An open system has matter, energy, and information exchanged with the environment. Thus, its internal state will not be in equilibrium, since "exchange" necessarily entails "flow" (flow in a general sense, not restricted to liquids), which again entails a driving force, simply referred to as "force" (in a general sense). It is actually a relevant physical quantity (state function) in the spatial distribution of the gradient with T as a state function and r_i as a space coordinates component. According to the relationship of "flow" and "force" which could be a linear relationship, i.e., $J_k = \sum_{i=1}^{n} \alpha_{ki} X_i$, where J_k is K flow, and X_i is the first driving "force". The occurrence of this case is generally not very "far" from the equilibrium state, which means X_i is not very enormous (in equilibrium state J, X is zero). Although the relationship between "flow" and "force" is complex, near the reference point (now zero) it is allowed a linear approximation with a expansion of series! This state is usually referred to as a non-equilibrium linear state. The physicist Onsager found in the non-equilibrium-state linear region a commutation theorem, that is, $\alpha_{ki} = \alpha_{ik}$ i.e., i "force" generates k "flow" with the same effect as k "power" generates i "flow".

Chapter 3 The Current Development of System Theory   43

In the non-equilibrium linear region, the spatial distribution of the system state function can be different but does not change over time, this state is called the non-equilibrium steady state. For example, concerning a metal bar with one end contracting boiling water and the other contacting running cold water, if the time is long enough, the temperature at different points of the bar will be not the same but does not change over time. According to Professor Prigogine's equation $\Delta S = \Delta S_i + \Delta S_e$, the rate of entropic change $\Delta S = 0$, which is when different points on the bar don't change in temperature, and the temperature gradient does not change either, that is, $d\frac{\partial T}{\partial r}/dt = 0$ but $\partial T/\partial r \neq 0$. The pole absorbs heat from the boiling water and dissolves this heat into the cold water, so there is entropic flow $X_i J_i \neq 0$, which does not change over time. At this time, there will be $X_i \cdot J_i = 6$ (space-time density that gives rise to entropy). As a matter of fact, there is the phenomenon of irreversible thermodynamics, which entails the increase of internal entropy $\Delta S_i > 0$. The exchange of energy with the outside world, ΔSe can be positive or negative. Here it is a negative ΔS (the rate of entropic increase is zero), that is, the entropy does not change, which verifies the principle proposed by Professor Prigogine that a near-equilibrium state produces a minimum amount of entropy.

3.2.2 The Principle of the Generation of New Order When Far Away from the Equilibrium

In the nonlinear state, between "force" and "flow", there is a complicated nonlinear relation, which is also far away from the high equilibrium. Professor Prigogine thinks that only an open system that is far from equilibrium can generate a new sequence because he thinks under the imbalanced state, a new sequence cannot be generated. This can be understood from two aspects: A new sequence can achieve order only by struggling through disorder. The struggle is reflected by a major increase and then a major decrease in entropy which results in the formation of a new sequence. But in the imbalanced linear state, entropy increases at a low rate, and it is difficult to "break", because of no "break" leads to no "set", therefore a new sequence will not appear. On the other hand, in a linear state, there is the only possibility of quantity change instead of quality change, so a new sequence will not be produced. Professor Prigogine thinks that only when under external influences the open system is far from equilibrium to the threshold that results in relevant mutations, making the system total entropy change $\Delta S < 0$ (eliminating the increasing tendency of the system entropy due to "landing"), which then will produce a new sequence. This far-from-equilibrium structure that can produce a new sequence is called the dissipative structure (dissipation of entropy) by Prigogine. The meaning of the "dissipative structure" is applied by some scholars not only in equilibrium conditions but also in imbalanced linear conditions. The dissipation of entropy's increasing trend is also called the dissipative structure.

3.2.3 Kinetic Process Model (Blusseletor)

When proposing the theory of dissipative entropy, Professor Prigogine also suggested a three molecular reaction kinetic model, which is called Blusseletor:

$$A \xrightarrow{k1} X$$

$$B + X \xrightarrow{k2} Y + D$$

$$2X + Y \xrightarrow{k3} 3X$$

$$X \xrightarrow{k4} E$$

◆ A and B are constantly consumed in the reaction but constantly replenished by the outside world

◆ D, E are the results and are taken away as soon as they are formed

◆ The concentrations of X and Y are state variables that are constantly changing

This model is widely used, which is characterized by dynamic self-organization, while the third type is a nonlinear autocatalytic link. $2X \to 3X$ (Y as the supporting force) plays an important role, and the arrow $\to$ has multiple meanings, such as the formation, evolution, and occasion..., "plus" symbol is also used in a general sense, such as support, adding for reaction, and supplementation. For example, basic knowledge A evolves to X, and with the support of B theory X evolves to theory Y and conclusion D (D can be considered as the result.) The expansion of X in combination with Y forms a complete theoretical system ($3X$). X could be put into force to generate E.

With an emphasis on the phase of qualitative change in research on the evolution of a system, many scholars engaged in this area have made some achievements, such as the positive feedback theory, mutation, and bifurcation theory. What is more, the research on the mutation phenomenon has a close relationship with that on the stability of the system (In theory, they are of the same conceptual strain.) which supplements and supports the studies on the stability of system. Concerning "mutation", a special branch of theory may arise. The above-mentioned content can be considered a detailed study of the important features of the self-organizing movement process.

The field of self-organizing kinetics is not yet a complete open kinetic system of development, because it inevitably involves complex nonlinear sciences, and is thus intimately connected with many cutting-edge sciences, such as chaos theory, fractal theory, and neural network theory. Some scholars also include it in the field of self-organization kinetics. Now it seems that the intertwining of different branches of various disciplines is the objective reflection of the comprehensive associations of things, and rigid fixation of the "division" of disciplines is unscientific and impossible. What is more important is to understand the regularity between them and the development of "relationships" to enrich our understanding and application.

3.3 Self-Organizing Evolution—Paradigm of Self-Organization

In his representative work, *The Self-organizing Universe: Scientific and Human Implications of the Emerging Paradigm of Evolution*, Jantsch Erich established the self-organizing evolution theory, deeming it as a unifying paradigm that can incorporate various evolutionary phenomena—a self-organizing paradigm.

The evolution of the universe is a kind of motion, in which process the time and space beam functions and structures of a motion system in evolution become more complex and advanced, i. e., the "motion" becomes more complicated, while this evolution in the universe is in keeping with the uniform paradigm: that is a self-organizing paradigm—rule-abiding self-organizational motion, the core of whose formation originates from the interactions of related things, constituting "relations" (dynamic). The author holds that the theory of Jantsch Erich is the application of the theory of dissipative self-organization on large scales, which meanwhile foregrounds the motion process formed in the interactions of things.

3.3.1 Dissipative Structure: The Necessary Basic Structural Mode of the Survival and Evolutionary System

The properties and concepts of the "dissipative structure" are proposed by the physicist Prigogine, which are opposed to the characteristics of isolated systems in that it has energy, material, and information exchanges with the external world. A dissipation means dissipative entropy, which constantly increases the entropy of the system. Dissipative structure is the most basic feature required in the development of systems.

Without input, there can be neither sustained "interaction" nor self-organizational dynamics. Therefore, an open dissipative structure is a basic structure for the survival and development of systems. In this structure, there can be other features on higher levels, such as self-sustaining power and various cyclical characteristics. The former refers to the self-renewal functions of a system, while the latter can be further divided into several categories in accordance with characteristics, which in term can be divided into sub-categories, forming a more complex environment. Circulation can be seen as a complex cycle in wide ranges.

3.3.2 Macro-Order of a System

This is an axiomatic law. To a living system, perceived from the level of entirety, its motion must obey certain rules, which are termed macro-order that forms the survival characteristics of the system. It is self-evident that if the motion of a system is disorganized and chaotic, its characteristics cannot stand. Macro-order also has another significance, which besides referring to the level of the whole system further designates the major sections of the motion features. Orderliness overall does not necessarily limit the dispersive fluctuations of its motion laws, with the sections of lesser significance having fewer limitations if only the order of major sections is not disturbed.

For instance, the macro-order of a middle-school class consists of teaching and learning activities, which must reach a certain standard at the same time (It is relative, with prestigious middle schools differing from ordinary ones, and classes differing from each other.) The other characteristics of the students can be diverse, but they mustn't influence the standards of classes.

It needs to be further emphasized that "macro-order" is not changeless, which develops with the motion of the system. It will be further elaborated on in the following chapters. Macro-order arises from the self-organizing laws as evinced in interactions.

3.3.3 Through Fluctuations the System Reaches a New Order: the Beginning of the New-Level and Process of the Evolution of the System

"Fluctuations" are the inevitable variations in a complex motion process, which causes can be summarized into two categories: the changes in outputs and the nonlinear internal characteristics and their feedback. The aftereffects of "fluctuations" are also related to the structures of the moving objects, e.g., possessing stable structures can facilitate the absorption of a certain degree of "fluctuations", which, if pushed beyond certain limits, however, will lead to the change of the order of the "motion". From the law of the unity of opposites, it can be deduced that "fluctuations" as the expression of the motion of opposites are ineluctable! As to human society, there is constant development in that the former structures of various social systems will undergo changes and even reconfigurations while their corresponding functions will expire to be replaced by the new functions of new structures, with the emergence of new structures, diverse sections, and related functions. All of these are incidental fluctuations that break the older. However, these incidental fluctuations have laws and directions, from which sense they can be deemed necessary. Thus, this law should be perceived as exemplifying the combination of "necessity" and "probability".

For example, wars are "surface fluctuations" in society. On one hand, large-scale wars would cause great damage; on the other hand, they are also "detergents", which can educate and unite the people as well as raise their awareness. For instance, Germany and France fought hundreds of years for the coal and iron resources in Ruhr and Saarland area. After World War II, people from the two countries (especially France) contemplated how to solve this problem. This see-saw type of competition was not the way out. When the thought of a joint development came to their mind, they created a Franco-German coal and iron union, which gradually developed into the European Community. In a similar manner, although the Chinese people paid the great cost in the Sino-Japanese War, it greatly tempered and educated the Chinese people, especially the liberated areas and the army led by the Chinese Communist Party, which laid a solid foundation for the founding of the People's Republic of China. However, every victory that gradually accumulated against the Japanese army was a combination of "necessity" and "contingency".

"To achieve a new order" is a process, which includes gradual mutations and sudden change. A sudden change is a qualitative change and forms based on a quantitative change. For example, the integration of the development of mobile communications into society as an integral part of society is composed of several qualitative and quantitative changes: the first qualitative change lies in the differentiation from the military load, vehicle load, and mobile radio, and in the establishment of portable systems (analog) of the commercial and mobile phones. The second qualitative change is the adoption of digital microelectronics; another qualitative change is the formation of unified protocols and standards, resulting in the formation of a global network. But each time there must be a quantitative change before the qualitative change, for example, if there is no quantitative development of analog, there will be no digital development needs, etc., while the next phase of development, is likely to integrate into the society. A greater network of information services is becoming an indispensable part, which is a negation of the simple function of an individual and isolated mobile phones!

3.3.4 The Co-Evolution of Macro-and Micro-Cosmos: the Common Evolution of the Environment and the System Itself

Darwinian theory of evolution is a great revolutionary scientific discovery, but it still carries the stigma of a conservative system, namely, it emphasizes passive selection on the microscopic level, and its spatial scope is relatively narrow, focusing only on single species instead of the actual analysis of the co-evolution of the existing systems and the environment in a larger range. The co-evolution of the macro-and the micro-cosmos is a good supplement, improvement, and development for Darwinian Theory. The law of the common development of the macro-and micro-cosmos is derived from the interaction of the overlapping sets of multi-level systems and multi-sections of regular self-organization.

Various laws and theories have their corresponding ranges of applications, which are necessarily limited to their effective areas, for instance, Newton's gravitational theorems can be applied to objects in the universe (including the outside of the solar system), but Darwin's theory of evolution, the author believes that: it is even not confirmed that there is life beyond earth, so it is appropriate that the movements of celestial objects are termed changes. Therefore, the co-evolution of the macro-and the micro-cosmos is mainly earth-based co-evolution of the human world, and it is further to be noted that: biological evolution itself happens in long temporal scales and complex spatial environments. Mankind has not yet grasped subtle rules and is trying to change characteristics of the living things through genes. The following discussion in this section is mainly about the evolution of the human environment (including the enrichment of required material) formed through the understanding and practices of human beings.

1) Basic Equivalence of "Evolution" and Symmetric Breaking

Symmetry: The antithesis to a reference system exists and their distance to the reference system is equal (Such as in human eyes, the nose is the center position on the symmetry (can be regarded as)) and does not change is called symmetric to a reference system. Like symmetry to a certain point, an axis, a plane, a space, and hyperspace. The symmetry to the timeline, in mathematics, is $F(x, t) \neq F(x, -t)$ regarded as a reference point in time; while in physics, it can be regarded as the reversed dynamic characteristics. The time asymmetric forms the asymmetric spatial. Most advanced complex movements are space-time asymmetric. Such as, life activities must be multi-level and space-time asymmetry. As "symmetry" is a movement characteristic of things, the "symmetric" feature of its opposite must exist. In a complex movement, there would be multi-level "asymmetric" "Symmetry" coexisting. For example, individual life within a national movement is asymmetric; however, the population of the whole nation will remain unchanged. Although a great cultural figure's life ends, his contribution to culture enjoys a certain immortality.

"Evolution": in its own sense is irreversible. Things are developed from the low-level, simple to advanced and complex. Its main feature is a qualitative change. Therefore, in the specific level it will be able to find the corresponding space-time symmetry breaking in the evolution.

In a broad sense "Evolution" includes the general process of the disappearance of specific things. Such as the biological extinction of some species, many of which are not suited for the evolutionary process. There is also environmental and ecological environment damage resulting from the incomplete or not profound understanding of systemic factors in the development of human evolution. But as the world evolved and human development, these old problems will be solved to some extent. However, there will be new demise in the process, and it can be seen as occurring asymmetry. "Asymmetric" in some sense can be seen as the incentive to development, which is another meaning of asymmetry. Because the "asymmetric" means a breaking of things at "balance", it is more difficult than to produce a new sequence (Law of Prigogine). For example, some physicists believe that the gradual evolution of matter and antimatter asymmetry forms the present world, and forms plants, animals which gradually evaluating to humans, and humans. The social development, such as the production of agricultural workers began to have a surplus in addition to supporting themselves, and because of this humans have further development (The lifestyle from hunting to agriculture is a big step because plants can use solar energy much more efficiently, the same land area can feed more people than before.)

The evolution of complex things usually involves cycling reactions. The following is an introduction to several "cycles" and their formations:

Reaction cycle: "cycle" refers to the cycling motion of food, and the reaction cycle has many sections, with the upper section supporting its motion forming a closed circle. For instance, the plants grow with the animals under sunshine forming the existential circle of organic beings, as shown in the Figure 3-1:

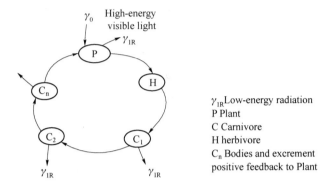

Figure 3-1 The Existential Circle of Organic Beings

In the reaction cycle, due to environmental transformations, expired (negated) sections are incorporated into cycles (the negation of the negation) forming new cycling mechanisms, which are called section cycles of the negation. For instance, Y-5 cargo-transport biplanes have been withdrawn from the transportation system, but they can be recycled in agriculture for pesticidal efforts to increase food production. In actual existential motion cycles, each cycle is dissipative and entropic, the same as the entire cycle.

Catalyzing cycle: referring to the cycle that has at least one intermediate which acts as the catalyzer (intensify effects). Catalyzing reactions consist of two types: self-catalyzing and interlocking catalyzing reactions, which will be tentatively explained by using the Brussels reaction of Professor Prigogine:

$$A \xrightarrow{k_1} X; \tag{1}$$

$$B + X \xrightarrow{k_2} Y + D; \tag{2}$$

$$2X + Y \xrightarrow{k_3} 3X; \tag{3}$$

$$X \xrightarrow{k_4} E; \tag{4}$$

Among them, X, Y are intermediates, and through X, Y, D, E are formed. (3) represents self-catalyzing reactions (to speed up the increase of X), (2) and (3) jointly form interlocking catalyzing reactions.

The entire process kinetic equation is:

$$\frac{dx}{dt} = k_1 A + k_3 x^2 y - k_2 Bx - k_4 x + D_1 \frac{\partial x}{\partial r}$$

$$\frac{dy}{dt} = Bx - x^2 y + D_2 \frac{\partial^2 y}{\partial r^2}$$

Among them, x, y are the values of X, Y respectively, D_1, D_2 are the diffusion coefficients of, x, y respectively, and r is the spatial coordinate. When $k_1 \ldots k_4 = 1$, D_1, D_2 is very large (diffusion is fast), $\frac{\partial^2 x}{\partial r^2}$, $\frac{\partial^2 y}{\partial r^2}$ can be considered as 0, and the above equation is simplified as:

$$\frac{dx}{dt} = A + x^2 y - Bx - x = f_1(x, y)$$

$$\frac{dy}{dt} = Bx - x^2 y = f_2(x, y)$$

It is still a nonlinear system of differential equations.

Depending on the parameters and starting values, the x, y can be of various types, such as Hopf bifurcations, limit rings and stable solutions, expressing to some extent the complexity of the motion, details of which can be found in the relevant research literature.

Complex super-cycle: a cycle that contains at least one catalyzing reaction section is termed a super-cycle. The human world is a more super-cycle system consisting of multi-sectional, dynamic, and complex super-cycles. Human beings' dwell in this system as well, the operations and dissipative entropy of which are concerned with the crucial existential problems of mankind, including the environmental problem. It also embodies numerous crucial problems that should be noted in the evolution of self-organizing entropic motion as well the essential theoretical effects.

2) Co-evolution of macro-and micro-cosmos, and survival the fittest

This principle applies for the evolutionary "cosmos" (the systems and the system of these systems constituted by evolutionary things) and to some extent, supplements Darwin's Theory of Evolution. It exhibits that evolutionary things are interrelated and interactive (They are mutually promotive, checking on and co-existing with each other.). The meanings of the "macro-and micro-cosmos" have no fixed sphere. To a concrete system, it is evolution based on an open and dissipative organization; to a larger system, it is evolution in an even larger open environment. The sphere, however, is not absolutely unfixed. I hold that the earth exists and evolves in the solar environment, while the

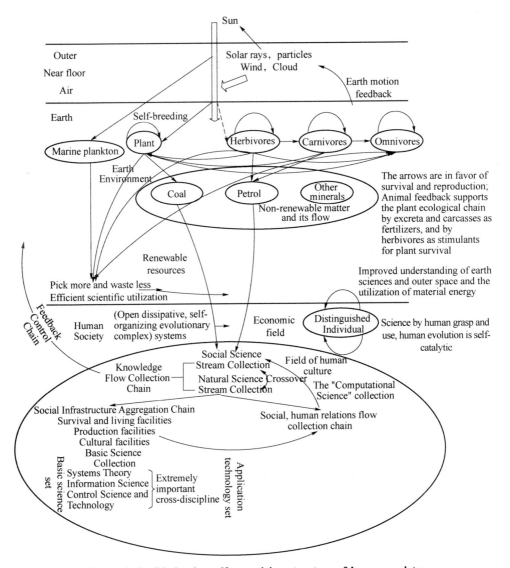

Figure 3-2 Dissipative self-organizing structure of human society

evolution of the solar system (for example, the Big Bang theory) calls for further exploration, or at least, it is difficult for ordinary people to understand. This relates to the fact that the applicative sphere of the dissipative self-organization theory has always been changing, but the movement of things on the earth conforms to the theory.

The human world exists in a giant open complex system in which human beings play a leading role (proposed by Mr. Qian Xuesen), but the consistent development of their cognitive and practical abilities must be guided by their continuous improvement of reason. The reason human beings should consist of their continual understanding of and respect for objective rules (including the understanding of and respect for this principle). Human beings have hurt the environment and upset the ecological balance with the excuse of various "needs". Severe environmental destruction will eventually exert a negative influence on human existence, which is a punishment resulting from disobeying the co-evolution of the cosmos. The human world (also a cosmos) is in the process of evolution and the development of the material culture (material wealth and environment) becomes progressively faster. The reality shows that human beings have observed the principle, while they need to enhance their consciousness. The principle also implies that the evolution of a system usually starts from a localized change and develops into a whole one. For instance, the economic reform of China initiates in the countryside of Anhui Province where a dozen farmers volunteered to institute the household contract responsibility system and later it was promoted to other regions of China as a nationwide reform. Shenzhen served as the experimental unit in the Reform and Opening up, which was popularized throughout the nation and at the same time, greatly promoted the development of Shenzhen. It should be pointed out that the principle is about the multilayer dissipative self-organization, playing an independent and interactive role in the evolutionary process of multilayered system.

3.3.5 The Evolution of the Evolutionary Process (More Applicable for the Evolution of Human Society)

The principle shows that evolutionary things on the whole have the evolution of the evolutionary process, which is quite important to an evolutionary human society. It reminds people that the mechanism of their long evolutionary process is changing rather than unchangeable. In the process, human beings should take advantage of the positive aspects and avoid the negative.

The reasons and ways of the evolution of the evolutionary process:

The pushing effect of the social culture of human beings includes the accumulation of material culture, the increasing of social wealth, and the enhancement of social development ability, which promote the evolution of the evolutionary process and the development of spiritual culture, as well as make the human society grasp and evolve related scientific rules and promote the development of evolution with them by increasing the efficiency of the subsystems' self-organization such as increasing the cycle and adding a new link in the catalytic reaction and forming new structures of complex supercycles.

The manifestations of the evolution of the evolutionary process:

There is a significant development in the overall evolution in the spheres of space and time, for example, the increase in important evolutionary nodes, such as the development of culture (including the development of education), productive forces (the development of productive relations), important social infrastructure and the quality of human life.

Sections in 3.3 introduce and explain the evolution of self-organization—self-organization paradigm (proposed by Professor Jantsch), but the author thinks that the importance of a theory concerning a complex problem is not in its rigorous and comprehensive discussion or detailed introduction, but in proving its function as a basis in the practice.

3.4 Professor Qian Xuesen's Propositions and Contributions

Professor Qian Xuesen is the first one to advocate the methods system which combines with the systems engineering in China's engineering practice which contributes greatly to the improvement of efficiency and quality of engineering practice, and the reduction of failure risks as well. He extracts the theory of systems engineering by integrating the theory of engineering control with the practice of China's guided missiles and spacecraft system and extends the theory to wider practical scopes. On this basis, Professor Qian Xuesen makes further contributions to system science: in his study on the system theory (systematics) which is the theoretical basis to the systems engineering, he thinks deeply of the "difficult points", i.e., the complexity, tracing back to the philosophy of dialectical materialism, and analyses and synthesizes the basic principles of "contradictions". His contributions are mainly embodied as follows:

3.4.1 The Proposition of the Concept of "Open Complex Giant System"

In the study of complex giant systems characterized by society, openness to the

Chapter 3 The Current Development of System Theory 55

outside world is another important feature, and thus the open complex giant system is proposed as the backbone system in systems theory. This is the opposite of the usual Western philosophical thinking, i. e., starting from the concrete, metaphysically discussing things, and thinking from the bottom up, which is to start from the global point and the first form the overall. This is a major systemic mode of thinking that starts with the big picture, form the framework and then extends to the details and feeds back into the big picture. The above refers to the whole process of knowledge, while the top-down mode of thinking refers mainly to the deeper stage of rational knowledge of complex things, forming a deeper and more comprehensive knowledge (systematic knowledge), while the whole process of knowledge is still from perceptual to rational, and again in a circular fashion to infinity. (The objective world, things interacting with each other, forming systems, forming complex systems, complex giant systems...) Professor Qian Xuesen believes that the further development of system theory should emphasize the open complex giant system as the backbone system, research, and analysis to constitute a theoretical system from the top down.

3.4.2 From Qualitative to Quantitative Synthesis—the Core Methods of Research on Giant Complex Open Systems

In the above analysis, the concept of a giant open complex system is proposed, which calls for corresponding methods for its research. After longtime deliberations and discussions, practices, and tests, Professor Qian Xuesen proposes the core method of "comprehensive synthesis from the qualitative to quantitative", which is also the core methodological framework, providing specific arrangements for specific problems.

First, we can perceive from its name that in the open complex giant system there are external and internal, upper, and down interlinings and interactions, which dynamically form very complicated spatial and temporal relations. If these are ignored, and problems are dealt with simplistically, key points and rules will not be grasped, which leads to "necessary" blindness, obstructing understanding, and progress. Therefore, a whole scale thinking framework is needed that is called qualitative thinking, which however is insufficient for accuracy. Therefore, meticulous quantitative thinking is supplemented. At the beginning of contemplating a problem, quantitative thinking will often end in bewilderment, or because of the immensity and numerousness of the problem, qualitative thinking is needed to provide a layout of the problems and dissect them into crucial ones, which are then turned over for the accurate resolutions by quantitative thinking, whose results are then tested in from the whole perspective as feedbacks. These two thinking modes and processes, together with the constant efforts of the experts are indispensable. The above-illustrated modes, processes, and key points are generally referred to as the methods of "comprehensive synthesis from the qualitative to the quantitative" (proposed by Professor Qian Xuesen).

It is the core concept in dealing with complex problems, the methodological framework, which has different contents and methods for the solution of different problems.

The synthesis from the qualitative to quantitative often begins with qualitative concepts, which undergo gradual intensification, proposing assumptions concerning the key points of problems (which can be qualitative as well as quantitative, or both). They generally resort to experiences (professional experiences), or the joint efforts of various experts, which are devoted to the analysis and testing of these assumptions, usually relying on reflections to form models for simulation testing.

The establishment of models and the modeling program, together with the understanding of the results of simulation testing is usually the joint efforts of various experts, which often undergo repetitions.

The advanced simulation testing method is to apply computing platforms in modeling and simulation for analysis and testing. The modeling process (the establishment of initial models, the simulations of mid-term results, and the modification of models tother with the analysis of results) calls for the constant involvement of experts, which consequently is a process that combines human beings and machines.

The acquisition of knowledge from qualitative to quantitative synthesis is a repeated process, not only with main cycles but also with embedded sub-cycles, which are in turn embedded with sub-sub-cycles with feedback processes, and constantly developed in practices.

Human society is a giant open complex system; the evolution of human beings now mainly manifests itself as social progress, which is dynamically constituted by almost innumerous interlinking subsystems, such as actual societies (the sub-societies of the whole society), the sub-societies of the actual societies, and so on...

The self-organizing mechanism consists of a reaction cycle, self-catalyzing cycle, super self-catalyzing cycle, and mutual feedback reaction evolution as well as other modes.

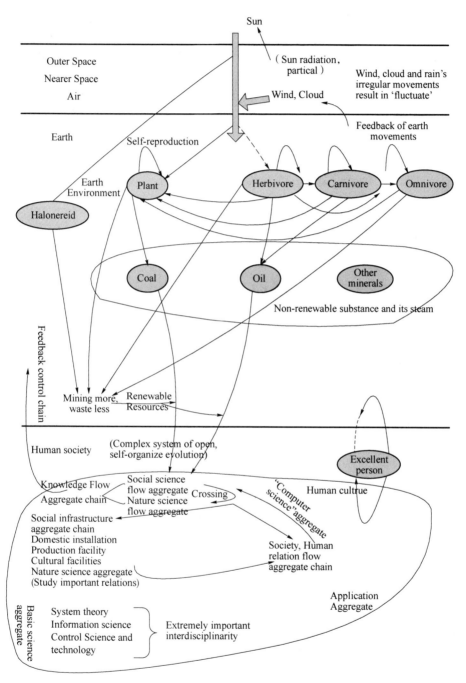

Figure 3-3 Demonstration of dissipative self-organizing structure and evolutionary relationships in human society

The dissipative mechanisms and modes of human society are diverse, but in research on dissipative mechanisms, it is of primal importance to restrict the emission of dissipative materials. For they can never be released to the outside of "the global village", it is important that they be recycled. Also important is the improvement of the efficiency of input flows. See Figure 3-4.

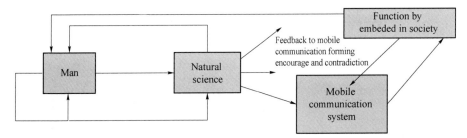

Figure 3-4 "Co-evolution " is in practice catalyzing reaction cycle with various interactions

Chapter 4
Information and Information Systems

4.1　Information

The word "information" has become popularized in society since the 1980s. Topics of people's talking are mostly around information or closely connected with it. But what is "information"? We can subtly understand its connotations but find it hard to put them into words. Early in Tang Dynasty, a poet named Li Zhong wrote a 7-character-per-line poem that says "I haven't heard the information of my friend for long, even in my dreams; leaning on the balcony, I wait for her coming". This poem is believed according to textual research to be the first historical record in the world in which the word "information" first appears. Chinese classical poetry is the condensation of "artistic conceptions" and multi-layer imagination, stressing "subtle appreciation" but not "description", so the meaning of the word "information" included in this poem is implicit. It's a great regret that even in the records and documents of the then history there was no explicit definition of information. Not until 1948 did professor Shannon defined and quantitatively describe "information" after referring to the connotation of entropy in thermodynamics.

4.1.1　Connotations and Definition of Information

Ancient Chinese used information but did not explain it. Likewise, in the past long history, people from other countries constantly applied "information" but didn't define it until the year of 1948 when American professor Shannon defined information as the elimination of uncertainty because of the process of gaining information via signal transfer in the field of communication. He borrowed from physics the concept of entropy to quantitatively express information as $I = -\sum_{i=1}^{n} p_i \log p_i$, $\sum_{i=1}^{n} p_i = 1$. This is a milestone in the development of information science and technology. With the advancement of human society, the connection between human beings and the field of information becomes closer and the scope wider, even beyond the limit of the communication field. Information has become one of the most frequently used words but still has no concrete definition. Actually, it's impossible to form a quantitative and all-inclusive definition of information currently; at least it's hard to do. Deep-going understandings and quantitative descriptions of the nature of information have to be carried on in the long run, and here we put forward for the time a qualitative generalized definition of information as follows:

Information is the representation and description of the state of motion of objective things, in which "representation" is the representation of objective existence, and "description" is artificial.

The significance of information lies in its capability of representing an objective existence, integrating with the practice of human knowledge, and then further integrating with human existence and development. So the advancement of science and technology in the information field reflects an important aspect of the integration of objectivity and human subjectivity.

For human beings, the most fundamental mechanism of "information acquisition" is mapping (with the help of mathematical language), i.e., the state of motion of objects is generalized and abstracted via human sensory functions and the brain's cognition functions, and eventually forms "knowledge". This is the process of "information acquisition" and "processing", a mapping from "objective existence" to human subjective knowledge.

Since the motion of objects is the dynamic expansion of the complex generalized space (not limited to 3 dimensions) and time dimensions, its representation is bound to be complicated, which is embodied by the intertwined development in the multi-stage and-time interval time dimension of the multi-layer and-section "relations" existing in the generalized space dimension. This further proves that information is constituted by the information segments which reflect each layer and section's motion states in different time intervals. This is the basic connotation of the inner structure of information.

4.1.2 Representation and Characteristics of Information

Information's objective representation is wide-ranging, for it originates from various motion states' characteristics, consequently, the representation of information is the reflection of various "specialties", which is also taken as the "expression of characteristics".

As for mankind, man can sense various related information from the nature using sensory organs and brain functions. Besides, man creates "symbols" to describe and record information, and then to study on the application of information via transmission and exchange. The above stated can be summarized as the process of dual mapping governed by the human brain, i. e., sensing and understating information to form concepts and thinking, then mapping them as symbols (In theory it can be explained in fractional steps, while in practice the thinking process of the human brain is integrated with "symbols". But the specifics of the thinking process of human brain are not clear.) In most cases, man also understands information through symbols.

"Symbol" is a word with wide connotations, and researches on symbols and their applications have already become a discipline—"symbology". Here are some examples of symbology: languages, scripts, graphs, and pictures, and disciplines including music, physics, chemistry, and mathematics. In these disciplines, besides languages and scripts, there are also specialized symbols, e.g., differential and integral symbols developing into symbols of operators, limits, norms, and inner products; and peculiar symbols like Wave-vector (State-vector) function in quantum physics as well. Given extended application, all theorems can be regarded as symbol sets—order construction of symbols, i. e., the generalized symbol, and also the symbol of objective laws. In addition, human expressions and motions (e.g., shaking head, waving hand, frowning) are also symbols.

4.1.3 "Information's" Characteristics and Summary of Its Characteristics Concerned

Characteristics of information's form of existence (direct layer)

The information is not matter, nor energy, but the representation and description of the state of motion are closely related to energy and matter. The motion and change of matters are constant, so information is not conservative.

Information can be duplicated on certain conditions according to the non-quantum action mechanism (in the quantum action mechanism information can't be accurately "cloned".) Idem the condition, "information" possesses the characteristic of multiplex.

We can understand it via the quantum energy representation equation in quantum mechanics: the mutual relations between things reflect that things function on and influence each other. To keep its state uninfluenced on the condition of mutual functioning, things need to have much higher energy than energy valve value and the energy change resulting from the mutual functioning. The numerical order of the lowest valve value equals $\varepsilon = h\nu$, i.e., in equation $\varepsilon = nh\nu$, when $n = 1$, the equation represents the lowest possessed energy in the existence of quanta, in which $h = 6.626 \times 10^{-34}$ joule second, ν the frequency range, assume $\nu = 10 \times 10^9$ second and $\varepsilon = 6.626 \times 10^{-24}$ joule in electronic microwave frequency range; assume $\nu = 10^{14} \sim 10^{15}$ second, and $\varepsilon = 6.626 \times 10^{-19}$ joule in light wave frequency; in microwave range and light wave range the energy levels of "operation" are $10^{13} \sim 10^{14}$ W(joule/second), and 100 photon energy levels, their energy levels are far higher than ε. In the meantime, the energy level that the information sources produce information is generally higher than the above-stated lowest operation level. So in the process of "operation", the self-state of "information" is stable and can be repeatedly duplicated.

As for that information whose forms of existence are represented by "labels", their applications depend on outside energy, e.g., scripts and graphic information are recorded on paper, and their information are evinced depending on lights reflection; the energy levels of their operations are also far higher than ε, and do not influence the state of matters loading information, so they can also be repeatedly utilized.

In the quantum field of matters, due to quantum's decorrelation nature, on certain conditions, quanta cannot be cloned ("untouchable"), i.e., they cannot be multiplied. This though strengthens the security of information; even weak interference in the environment will break its original state, which entails great difficulties in normal utilization (These problems are still under exploration.)

The characteristic of sharing: when the information carrier has energy or when its equivalent energy is far greater than ε, environmental interference energy and the energy necessary for information utilization, then this information can be shared in multiple places. E.g., voices can be heard simultaneously by several people; multiple receiving stations can simultaneously receive symbols and obtain information through satellite rebroadcasting. The 3 characteristics of information discussed above convenience mankind in utilizing information, but at the same time impose difficulties to the work of information security. Research on the reverse utilization of the above characteristics are carried out, e.g., researches on quantum code (far from practical use), and open up new research directions and fields.

The time dimension characteristics of information: The specific motion of objects is always conducted in the limited time and space dimensions. Consequently, information is bound to have time dimension characteristics, e.g., the time of happening, the longer it lasts, the time intervals, the rate of change to time, and the mutual sequential relations, these are the prominent characteristics in time dimensions in "information's existence form", and of great significance to the utilization of information.

The 2nd layer—the characteristics in the layer of information utilization concerned by mankind.

The most fundamental and essential function of information is its utilization by mankind, i.e., being utilized with a man as the subject; therefore it has the following characteristics:

Facticity: the origin that information doesn't truly reflect the state of motion can be classified into two kinds: intentional and accidental. Accidental means the distortion of information is caused by the fault of man or information systems, while intentional refers to a man deliberately producing false information or altering the content of information to achieve a certain purpose.

Multi-layer and section discrimination characteristics: to which layer and section "information" belongs is also an important property. For information on complex motions, it's necessary to know their layer and section properties to have a comprehensive grasp of the information of motion.

Selectivity of information utilization: Information is the representation of the state of motion of objects, while motion involves various complex mutual relations and assumes image nature (subject, object, correlation...), i.e., in specific situations the correlation property of the content of information produces importance of different degrees to certain subjects, objects and correlations, e.g., it's of no great significance to that unimportant information. This property is called the selectivity of information.

Additional meanings of information: As the representation of the state of motion of objects, information may be just of certain section but necessarily includes complex relations mutually related in "motion". Obtaining the connotative but not explicitly expressed content (obtaining the additional significance) via information is of great application significance. Man obtains additional meanings through two methods: imagination and logic inferences, in which imagination as one of man's thinking functions (The mechanism of proceeding from one to the other is complicated.) is more widely applied than logic inferences.

For example, associating the nature of research with the new commodity the enterprise will create; employing logic inferences to determine the denotation according to the multiple information implying the denotation.

The motion of objects is "objectively existent" and has numerous complicated varieties. In the virtue that the deep-going significance of information lies in understanding the process of motion via the state represented by information, there are two key aspects in the characteristics of the "process" of man correlating "information":

"Information" doesn't omit to represent the key state in the process of motion, and information contains the essential factors that enable "state" to key the "process" of motion. Understanding the process of moving through the individual state (information) follows the process from part to whole (the process from knowing nothing to knowing something), therefore it is impossible to know in the state of knowing nothing (This is apparently a paradox.) So what we are concerned about is that contained in each piece of information are the factors representing the whole for us to "excavate". The schematic framework of excavating information's connotations is of the tetrad relation:

"information" => [information direct characteristic domain, the generalized space domain in which information exists, the time domain in which information exists and the information change rate domain].

Due to the complex variety of motion, the above-stated domains need further dividing into sub-domains for studies: information direct characteristic domain correlates with the following sub-domains:

Correlated subject sub-domain: e.g., things, matters, human beings, and the integrated sub-domains including that of human beings and things, things, and matters, and human beings and maters, etc.

Correlated conduct sub-domain: e.g., movement, desire, evaluation, judgment, and etc.

Property sub-domain of the state of motion: certainty, uncertainty (probability and non-probability uncertainty), associative property of certainty and uncertainty.

The generalized space domain in which information exists correlates with 3 dimensions of distance space sub-domains: "laws of things" space sub-domain, "laws of affairs" space sub-domain, "laws about humans" space sub-domain, and "laws of beings" space sub-domain. Those sub-domains can be further divided into multi-layer sub-domains and their characteristics further analyzed, e.g., "principles of matters" space domain can be further divided into sub-sub-domains including mathematics space, physics domain, chemistry domain, etc.

The time domain in which information exists needs to be divided into time domains of various scales:

Information change rate domain can be further divided into the following sub-domains:

Generalized space multi-layer change rate sub-domain:
$$\frac{\partial}{\partial x}, \frac{\partial}{\partial y}, \cdots, \frac{\partial}{\partial \theta}, \frac{\partial}{\partial r}, \cdots, \frac{\partial^2}{\partial x^2}, \frac{\partial^2}{\partial y^2}, \frac{\partial^3}{\partial x^3}, \cdots$$

Time domain multi-layer change rate sub-domain: $\frac{\partial}{\partial t}, \frac{\partial^2}{\partial t^2}, \frac{\partial^3}{\partial t^3}, \cdots$

Time and space multi-layer change sub-domain: $\frac{\partial}{\partial x}, \frac{\partial}{\partial t}, \frac{\partial^2}{\partial t \partial x}, \cdots$

For example, whether the movement process is in the qualitative phase to suppress the quantitative process, whether there will be a major new event generated, and whether the movement process is complex.

Some predictions about the process of motion represented by information can be made by using the above-introduced tetrad relations framework to analyze information (including integrating information) with the help of analogy and association.

Information aggregation constituted by information (information works)

The characterization of a state is represented by information, the information content (not penetrating facticity or falseness, importance and time property, etc.) included which can be expressed by baud and bit defined by Professor Shannon. But these are the information segments characterizing simple states, called "information units". In the objective world there exist information aggregations organically constituted by information units, which characterize more complicated motion states and processes, and are the natural extension of "information word". Yet they have no specialized name, here we call it for the time "information works" similar to what in Chinese semantics called "verbal works". Information works can describe and characterize more complicated motions by integrating logical inferences and judgments and are of great significance to the development of human society. The manifestations include script, picture and multimedia audio and video systems, etc.

Information and information media sensed by human beings

Man sense information via human sensory organs such as eyes, ears, tongue, and fingers and then transfers the information to the brain and nervous contrails for understanding. Man can sense altogether 7 types of information, and the intermediary part transferring the 7 types of information is called medium (Each type doesn't represent each kind.) And for its multiplicity, it is called multimedia for short, see

Table 4-1 does details. The above discussed refers to the media transferring information sensed by man. In the objective world there exist many other media transferring information, some of which are beyond man's sensory scope, and have to be "transformed" for man to sense, such as electromagnetic waves (Visible lights are not included), and ultrasonic waves.

After acquiring information, man can repeat and transfer it through an organization (most of the information is visual and auditory ones, also including integrated dynamic information and other kinds of information), and in this process man's subject, "description" is always added to the original. The one who describes should make sure of the facticity and objectivity of his description and the recipient should be able to extract through analysis the true and useful information and eliminate the impacts imposed by additional false information as well.

Table 4-1 Types of information sensible to man and the media

Visual information	70%-75% of the total information	Key media: script, graph, picture (visible lights reflection)
Auditory information	10%-15% of the total	Key media: auditory waves of 20-15thousand Hz (The visual and auditory information account for most of or even the overwhelming majority of the information sensible to man.)
Tactile information Gustatory information Olfactory information	20% of the total	The man mainly uses hands and skin to sense the physical states, the representative information includes softness and hardness, coldness and hotness, temperatures, etc. Gustatory information is acquired mainly depending on the tongue and the gustatory nervous system. Olfactory information is acquired mainly depending on the nose and olfactory nerves.
integrated dynamic information	No detail statistics, but its importance should be paid attention to.	This type of information is the organic dynamic integration of the above information, referring to the integrated deep-level information with the additional meanings that each sort of information doesn't include and that are organically formed in the human brain after man's dynamically sensing the sort of information.
Interactive integrated information	No detail statistics, but its importance should be paid attention to.	It refers to the deep-level information acquired on the bases of integrated dynamic information in the interaction between men, e.g., in academic discussions new senses sprout, and in research lectures both the teacher and the students can acquire this kind of information.

4.1.4 The Progress of Mankind Transferring and Utilizing Information

When man evolved into a primitive society, he had already possessed primitive sociality which manifested in the formation of primitive human society, originally for the purpose of collective hunting. One of the essential factors constituting society is mankind mutually transferring information to understand and think with the help of motion communication. Starting from sounds, motions and expressions used to communicate senses and intentions, through many ten thousand years of human evolution and development, language is formed (regular fixed phonetic sequences), then after another very long process scripts emerge (starting from keeping records by tying knots and drawing graphs). In the above processes, the scope of communication extends from near to far, and from small to large, as seen in Figure 4-1:

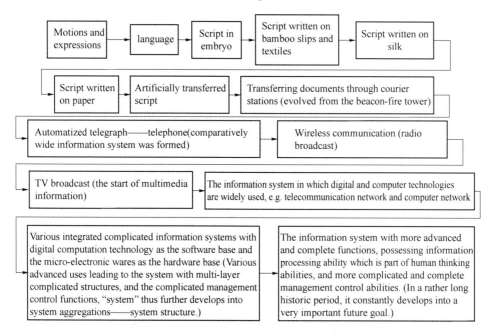

Figure 4-1 Mankind constantly perfecting information functions and the functions of utilizing information

There are three points that need further explanation about the chart above: First, in the process of human evolution, till the current stage, the process of utilizing information changes from slow to fast, and becomes more accelerated. From the primitive state to the formation of language and script, human history has experienced countless ten thousand years; the evolution from script to telegraph undergoes four to five thousand years, the time interval is greatly shortened and can be calculated by "millennium"; from telegraph to the large-scale information system using computer implantation, about 200 years; over recent 40 years, the process has developed extremely fast, and it's now still in fast development. Second, transferring and utilizing information. The Transfer and utilization of information start from simple media instruments; now information can be transferred and then managed and applied with the aid of modernized science and technology, and the "information system" used by man is gradually formed. "Information system", an important symbol of human progress, is not at all a simple instrument but an advanced artificial one. It's not until the recent hundreds of years, especially recent 40 to 50 years, that inventions and utilization integrating different systematic complicated instruments stand out.

Third, what are the development characteristics of information science and technology, and the information system in the new century and new millennium? Here is the answer: the development of information systems will be combined with the comparatively complete understanding and decipherment of the fundamental questions faced by mankind at present; the questions are: the understanding and solution of complex nonlinear problems, the nature of life occurrence (the qualitative change from non-life to life) and the key details in life continuation and evolution, human brain cognition science, and other challenging questions. Breakthroughs to the above questions need a historic period far longer than 20 or 30 years.

It appears that in the field of information science and technology mankind has developed into a prophase of a new historic period. We now thus come to the proposition of the development of information systems, main points of information systems are discussed as follows.

4.2 Information Systems

4.2.1 Definition of Information Systems

Systems that help people to acquire, transfer, process and utilize information are defined as the information system, which is an instruments supplying man with "information" services. The word "service" now contains wider connotations, so the information system is actually a generalized name of systems with various functions and characteristics.

4.2.2 The Systematic Theoretic Characteristics of "Information Systems"

Modern information systems are typical of the system defined by the system theory in which there exist many sub-systems overlapping and intertwining with one another. E.g., Modern communication systems consist of satellite communication systems, public backbone communication networks, mobile communication networks, and etc. Among which the satellite communication system is further divided into satellites (which consists of transponders, satellite attitude control, solar energy battery system, and so on), ground central station system (including ground control sub-system, up channel transmitting-receiving system, etc.), small scale user ground stations (which can be further divided into sub-systems). Mobile communication networks and public backbone communication network systems also consist of multi-layer sub-systems.

In each information system, partial improvements (quantitative stage) are continued even after the finish of research and development; when improvements fail to adapt to the situation, new types will be developed (a kind of qualitative change). This circulates and when it reaches a certain degree, greater structural qualitative changes will take place (systematic change in the system), such as the change in the switchboard in communication systems into stored program control. Now "the stored program control" is being changed into "route", which is also a systematic change. The endlessness of this kind of change accords with the principle of achieving new order through fluctuation in the system theory.

Information systems, systems serving human society and human beings, expand with the development of society and follow a distinct trend of co-evolving with society into the more complicated and advanced.

Yet the existence and development of each information system have their restraints. Furthermore, new developments bring new restraints and produce new contradictions as well. If the improvement of performance can be regarded as an acquisition, the acquisition has to be paid a high price.

4.2.3 Functional Constitution of Information Systems

Brief description of information system sub-functions.

First of all, extending the definition of information system in subsection 4.2.1:

"information" to serve human beings a connotation of tools, "service to humanity" has a multi-level connotation, here the focus on the direct function of information systems and from the direct level of upgrading and upgrading to the associated social development. The direct function of the information system and its elevation from the direct level to the social development level, each of which includes several sub-levels, all of which involve human-machine integration, such as human choice decision-making and control, are shown below.

The direct functional level of the system (elevated from the bottom to the top table sub-levels) is shown as follows.

Speculation on the deeper "purpose set" of the curated "ontology" movement (final curator purpose)

Speculation to determine the dominant self-organization mechanism of the "ontology" movement process

Perception of the main processes of the "ontological" movement and their characteristics

Perception of "information" corresponding to the main set of states of the movement of the "ontology"

The functions at the level of social development (from the bottom to the top of the same level)

Promoting the mastery of complex movements in the service of human social development

Coping with the impact (especially the negative impact) caused by the Movement

Using "movements" (for development progress)

Awareness of the Movement

In terms of the current level of information development, the use of information technology by human beings is still at a low level, and its direct functions are generally at the first or second level, resulting in a lower level of service to society.

All information systems are constituted with the following sections intertwined or intertwined selectively.

All information systems are constituted with the following sections intertwined or intertwined selectively.

The information acquisition section (various sensors). All information systems need one or multiple media to load information to operate to give play to the systems' functions as an instrument. So first certain medium is in need of sensitively acquiring "information" and record it if necessary. This is the essential functional part of information systems. It should be noticed that man constantly improves information depending on science and technology to acquire its partial performance and creates new types of information as well. The important breakthrough of acquiring devices and simultaneously information-acquiring parts of science and technology will certainly exert great influences on the development of human society.

In information storage section (including current semi-conductor memory, discs, etc.), information often exists within limited time intervals, so for the purpose of repeatedly using information, we need to store it via multiple methods, in which key performance indexes are required of rapidity, convenience, fidelity, large capacity, and multiplexity.

Information's transmission section (wireless information channel and sonic channel, optical cable information channel such as antennas, receiving and sending equipment). In this section, technology improvement in scientific studies targets large capacity, reduction of loss and interference, stability, and low prices.

Information exchange section (e.g., various exchange machines, routers, servicers). This section aims at the reduction of time delay, facility of control, security, large capacity, and compatibility of multiple signals and services.

Like the information acquisition section, these sections are now in constant development, among which the major developments exert great influences on human advancement.

Information's exchange handling section (Including various "multiple connections", signal encoding and decoding, modem, signal compression, depression, information measurement, etc., these all belong to the signal handling field.). This section is regarded as the bottleneck in the development of information science and technology, though having gained great improvement over recent years, still having no like-human information handling abilities necessary to the development to realize more closer link between man and machine. The way to realize this link is long and arduous, but still is one of the goals pursued by man.

System service support function: The main function of advanced information system is embedded in the society for service, service society has a variety of occasions and a variety of functions to form a variety of requirements for information systems, and the development of advanced information systems need to spend huge capital costs, so the design of information systems have a basic framework structure, and by the service support part of the flexible hardware and software configuration and more flexible configuration has met a variety of embedded service requirements, prompting the rapid development of "service support part".

Information management and control section (e.g., monitoring, valuation, fault measurement, contingency measurements under the condition of fault, multiple information business management, etc.) The finish of this section's function is becoming more complex and difficult with the flash increasing need for it resulting from the complexity of information systems (E.g., information systems' complex topological structures entails mathematic headaches on the science and technology base of the management and monitoring field.) With the information system further melting into society, social sciences enter, blend with, and further synthesize its discipline base of management control, and the complicated contents of social sciences and the humanities are thus included in its management control functions. This leads to the growing distance and contradictions between the "need" and the "practical level". E.g., the management control of e-business systems touches upon the field of law; multi-media literature and art involve management, ethics, and morality, law, etc. Consequently, the development of information's management and control section relates to numerous disciplines, imposing great importance as well as challenges and urgency.

All the sections possess the following characteristics: software combined with hardware, the discrete digital combined with the continuous simulation, and all functions partially intertwined and integrated to support the formation of the major functions section, e.g. the memory section are inclusive of the handling section; and the management and control section inclusive of memory and handling sections. The developments of the above sections are closely related to the new discoveries and the technological innovations in the field of sciences, and promote the mutual development between information science and systems, and human society. Challenges and opportunities suffuse the process of "development".

4.2.4 Basics of the Development of Information Systems

(1) Man's perception, understanding, and utilization of multi-media information—a conundrum awaiting study.

Man perceives and receives the information loaded by numerous media via the organs of the eye, ear, nose, tongue, etc., and man's understanding of things comes

Chapter 4　Information and Information Systems

from the integration of the information received and the brain's thinking. In general, man all has the possession-"in-itself" of the seven information handling abilities as included in Chart 2.1 in 2.2.3. But their detail mechanisms are rather difficult that even man does not understand himself. E.g., human eyeball acts as a lens to form inverted images, but the images man perceives via optical nerves are indeed erect. The mechanism involved is still not clear to man, at least not to a majority of people. The process of understanding and utilizing "integrated dynamic information" and "interactive integrated information" is even more intricate, and knowledge of it is extremely limited. In this situation, man naturally cannot produce high-level instruments concerning this section. Therefore knowledge and utilization of the information systems of the information loaded by multimedia have been man's "aspiration" and are still beyond man's reach. This kind of information system belongs to the information handling systems which are more complicated than the handling systems of individual information. The purpose of proposing this difficult problem in this unit is to explain the journey to develop information science and technology is endless and needs man's efforts.

(2) The development of man's utilization of electromagnetic media and the relative discussion.

See Table 4-2. Man perceives and understands the 7 information discussed above via sensory organs and the brain, which is the basic and cannot be changed. But man constantly expands in time and space dimensions the scope of perceiving and understanding information with the help of information exchange and information systems. The major measure taken to extend the scope of perception for man perceiving and eventually changing the information into the 7 types of information mentioned above, is certainly using electromagnetic media. The basic reason for this is related to the theory of duality of matter's wave granule proposed by professor Louis Victor de Broglie which claims that electromagnetic wave is the basic property of matter's existence. Actually, man is using sonic waves and other oscillating waves and gravitational waves (e.g., minute gravity) as well. In the near future, the correlation of quantum's tangling state can be used to represent the state of motion, etc. Besides, attentions have been paid to material information being used to represent the state of motion (e.g., fossil, rock pith, ice layer and loess layer) and bio-information. In conclusion, man will work hard to expand the time and space scopes in perceiving and understanding information.

Table 4-2 Development and use of electromagnetic media

Characteristics of wave length	Overlength waves	Broadcasting wave band	Short wave band	Ultrashort wave band	Microwave band	Millimtre wave band
Major developments (20th century)	1970's	20	30	40-	40-	After the 1970s—20th century
Modern uses	Communication in deep water	Broadcasting	Communication	Communication radar, remote control, etc.		
Characteristics of wave length	Long wave infrared wave band	Short infrared	Visible light	Ultraviolet	X ray	γ ray
Major developments	After 1970's	1960—1970	Two century's constant development	1980—1990	1920—1930	1980—
Modern uses	Night vision application, etc.	Space development application, etc.	Various new applications	Giving priority to special and military applications	Studies on matters structure	Studies on universe and ucleus

The generalized, as well as the important applications of electromagnetism, are up to 10^{20-25} in the frequency and scale scope, involving the macro effects (waves), micro and mesoscopic effects (particles and quanta) of physics, chemistry, and biology, the general direction and goal of which are developing after the ultimate goal stated later on. E.g., the satellite communication of ultrashort waves+ microwaves+ millimeter waves has been constantly improving and developing in the aspects of "anyone", "any place", "any time"; Light communication tends to the problem of speed transmission of large-capacity information resulting from "anyone + any thing", with light memory's large capacity being driven by "any thing"; Mobile communication is comparatively convenient and cheaper which tends to any state (mobile state). The developing functions of the internet: using computers (especially small-sized computers) to transfer and use information, but handling and using information are still a bottleneck. I personally believe starting from the 21st century, based on studies on life sciences, and brain sciences (cognition and thinking mechanism), man will energetically study information handling systems that can better cooperate with the human brain.

Information science and technology will develop along with the evolution of human beings. As stated above, it has reached a motion scale of 1020-25, about 1044 distant from the motion scale man has already mastered in nature in macroscopic view. As for the lower limit, it's based on the theory of "mutual functioning in matters" (which proves the complicated nonlinear mutual functions and extends in time and space), and its minimum quantitative mutual functions are still at a distance from the energy valve value. We thus can conclude that man's exercises in using media that load information and expand the scope of information acquisition will be in dynamic development.

(3) Eternal contradictions and the positive and negative effects.

The development of information science and technology and systems co-evolves with human advancement which is an open complex giant system, a self-organizing evolution system, which can further be divided into evolutions correlating with open complex giant systems, e.g., in the development and improvement of a region, or a nation, some aspects of information science and technology, and information systems will participate in the specific development as "special" forms. In this part, we will discuss the general concepts of eternal contradictions and their relevant positive and negative effects on information systems.

Human society's evolution results from the motion of contradictions which follows the law of the unity of opposites inside itself. When analyzed from the perspective of contradiction, it can be found that the closer the relationship between information systems and society, the more obvious the mutual promotion between them whereas the fiercer the struggles of contradictions. This is because the motion of contradictions is the origin of the development of things. All contradictions in society can be reflected in information science and technology and the information field, the specific representations of which can be summarized as the contradictions following the law of the unity of opposites, including absolutely and relativity, part and whole, quality and quantity, necessity and accidentality, continuation and discontinuation, etc. E.g., when reflected in development the contradictions between quality and quantity are represented at present by the "appeal" to the development of quality, but not quantity, expecting building links with human thinking via the development of "quality".

Part 4.2.5 which tells the ultimate limit goal of the development of information systems also largely involves the domain of the unity of opposites: the content of the unity of opposites is the source of the formation of the ultimate limit and the continuous development; information systems integrate with society and supply services, yet the great concerns of the information security contradictorily increase its severity and complexity, and the intrinsic dialectical contradictions of complicated matters. E.g., in the operation of complicated information systems, the interface for adjustment and test should be set up in advance (because the operation of the system may not be absolutely correct and may have faults), yet this interface may be used by the attacker as the entrance to implement attacks. The information systems serving the public are of course open to the public, the openness can be used to conduct criminal activities (e.g., stealing money via e-business), etc.

When the opposite contradictions intensify, confronting motions will take place, the functional effects of which therefore display mutually positive and negative characteristics. E.g., the two parties involved in the war are in the intensified confronting positions and thus take confronting actions. When this situation reflects in the information systems field of the military field, nations attach great importance to information security and the problems in confronting fields. The basic proposition of "eternal contradictions in information systems and their positive and negative effects" has been discussed in this part, and information security and confrontation will be discussed in the following chapters. Our generalized conclusions are: the eternal and universally existent contradictions of the unity of opposites and the struggles between their positive and negative effects are the source of the development of information science and technology and systems.

4.2.5 The Ultimate Limit Goal of the Development of Information Systems

In the past, the ultimate limit goal of the development of information systems is: anyone in any place at any time in any situation can securely and conveniently acquire information and use information (The cheaper, the better.) But this is not a scientific formulation, for the word "any" has absolute connotations and equals "without exception", which thus denies the specialty and contradictoriness of objective existence. In light of the above, the general goal of information systems should be: under the condition of abiding by social orders, energetically expand the scope of users and the use of users as well as reduce or weaken the restraints to the use in time and space dimensions to convenience users to obtain and use information (The cheaper, the better.) It should be noticed that when information systems are implanted into society to supply services, there exist contradictions between the "introduction" of social contradictions and the inner part of the information systems. It is these contradictions that promote the continuous development of information systems. Basic categories of contradictions of unity of opposites, e.g., relativity and absolutely, part and whole, freedom and restraint, finite and infinite, are constantly expressed in information systems, and the struggles of contradictions can be constantly found in time and space dimensions as well, e. g., the continuous expansion of functions and services and security. The improvement of science and technology simultaneously promotes the level of information security and the level of attack, which results in the fact that information security has no absolute safety but relative guarantee, the struggles between "security measurements" and anti-security measurements will be endless, and can only be found in dynamic development.

4.3 Development of Information Science and Technology and Information Systems: One of the Eternal Themes for Human Beings

4.3.1 Theoretical Discussions

The most essential factor in human evolution is the increase in the ability to learn and practice. "Information", as the medium, is crucial for the realization of the functions of learning and practice and the improvements of abilities, including the development of man's knowledge of himself.

For modern men, the major approaches in evolution based on knowledge and practice are the discovery of new laws, accumulation and application of knowledge, the invention of new tools, and improvement of tools. The acquisition of information is the most fundamental prerequisite for the above-mentioned evolution approaches. At the same time, various types of information systems are the most important tools for human beings.

4.3.2 Information Science and Technology and Systems as Universal "Enhancer" and "Catalyst"

As is described in the last section, "information" is a fundamental factor for the improvement of the ability to learn and practice, and information systems are an important tool for human beings. These two basic laws, together with the rapid development in digital electronic technologies and microelectronic technologies, enable many types of information systems to emerge in a smaller size, lighter weight, lower voltage, and lower power and thus facilitate information systems to integrate with other systems or embed in other systems as subsystems which play the role of enhancers. As a working mode, the information systems embedded into other systems as enhancers are of vital importance and are widely used; and this is also an important reason that makes modern society an information society. Examples are abundant, just to name a few: in the current civil aircraft, the navigation system, and the blind landing system are both typical information systems; they are embedded within the civil aircraft system to form modern advanced airliners, which are of high flight safety and landing safety, contributing significantly to the development of social transportation with quality service for passengers.

In the high-performance mechanical processing center, the embedment of relevant information systems is a key measure to improve processing accuracy and efficiency. Examples of this kind are ample.

In addition to the various information systems embedded in other systems, there is a different type of large information system which enhances and facilitates social development: it's a network structure widely distributed in space, having a multi-layer structure and operating with multiple scales in the time dimension. It uses common basic information infrastructure to build information systems for a wide range of information services; it obviously enhances and facilitates social development. The operating computer network and the various service systems constructed on this basis (including the national defense network system), can be considered the first generation of such systems. The next generation of networks being developed centering the "Pan-Existence" concept can be considered as a new generation of systems. The connotations of "Pan-Existence" can be considered as the exchanges of information and control of behaviors (or the Internet of Things) between man and man, man and objects, objects and objects, man and things, objects and things, and things and things. The development of this kind of information system has just started. It will be a rather long process along with the social development, resulting from the interwovenness and fighting of a variety of social conflicts (including all kinds of security issues); the basis of its development needs supports from natural sciences and technologies as well as that from the disciplines of the field of social sciences (such as law, sociology, ethics, and other disciplines).

4.3.3 Information Systems Offer Services in a Wide Range

They can be classified into the following according to their usage:

Research Information Systems (including the frontier information systems studying the life sciences)

Social Development Information Systems: weather information systems, disaster information systems; earthquakes, forest fires, floods; medical emergency treatment, epidemic detection, environmental protection testing, educational information systems, etc.

National Security Government Management Information Systems: national defense information systems; satellites, communication satellites, radar, and sonar;

Government anti-smuggling office management information systems: taxation and state administrative management information systems.

Social public services information systems: communication, computer network, radio broadcasting and television, cable television systems (simultaneously having the functions of publicity, education, and propagation in national organizations), public information banks, etc.

Economic development and enterprises marketing information systems: MIS, banking information systems, futures information systems applied in the stock market, e-commerce systems, and marketing and sales service systems.

Home Personal Information Systems: consumption and entertainment, and life and safety information systems.

Embedded information systems: refer to the information and information control systems embedded in other systems as subsystems to significantly improve and enhance the function of the original systems. For example, the navigation, autopilot, ILS landing systems in civil aircraft, the fuel injection control system, collision avoidance systems, road state, and navigation systems embedded in GPS in vehicles. It should also be noted that here "other systems" have broad implications, including a very wide range of complex and far-reaching social systems. For example, when integrated with modern information science and technology, they can construct artistic creations and they also can be integrated into all levels of the educational system.

Among these, the A-F information systems often share with each other the information infrastructure platform on which their specialized functions are finished via adding dedicated software and hardware equipment.

The service functions of information systems are integrating with each other on the basis of essential divisions, and share the infrastructure in the section of structure and composition to form systems with integrated structures yet independent key functions. For example, the long-distance service of mobile communication systems utilizes the network infrastructures found everywhere in telecommunications; e-commerce utilizes the computer network, whereas the computer network is inseparable from the telecommunication network.

In short, the development and progress of social civilization are closely related to the development of information science and technology and numerous information systems. Information science and technology, and information systems (including operation services) have developed into industries in themselves and at the same time, extensively integrate with and promote the development of other industries, thus realizing the interaction of developments in the field of economy and even culture.

4.3.4 Exemplifications of Electronic and Optoelectronic Information Systems in Vehicles (Huge Series of Products and Industry Groups Are Already Formed)

EMC System Design (Software System)

Vehicle Information and Control Systems: temperature control subsystem, pressure control subsystem, engine rotating speed control system, vehicle speed and acceleration (including angle dimension) control subsystem, flow control subsystem; over thirty types of sensors (with fiber-optical sensors) and bus systems (about 1000 signal lines without bus), not including auxiliary control subsystems, such as automatic lock, trunk, air conditioning subsystems, etc.

Service Safety Systems: reversing radar, collision avoidance radar (75GC), ABS (Anti-lock Braking System), airbags and so on.

Communication and Navigation Systems (Including Network Connections): electronic maps, road condition and electronic navigation and so on.

Environmental Protection Integration System: emission control subsystem, noise control subsystem.

Entertainment System: audio and video entertainment systems.

Development and Support Systems: software is due mainly in information acquisition, information processing, information utilization and formation of control; hardware mainly refers to the development and application of dedicated DSP, memory, programmable devices and ASIC.

Various Special Devices: special display, special sensors, etc.

4.4 Huge Support Systems for the Development of Information Science and Technology and Information Systems

As an important tool to human beings, information systems will continue to develop and evolve with human development. Its development relies mainly on the support of science and technology—the support from the multiple layers of "science and technology" intertwining and integrating with the relevant disciplines. Judging from the overall level of society, its development is reflected in the dialectical co-evolution of the information society, and information science and technology (embodiment of the principle of "concerted evolutions of universes of all sizes; survival of the fittest").

4.4.1 Discipline Backup Systems

Basic level: natural philosophy, language semantics, mathematics, physics, chemistry, biology (relevant sections).

Basic level of frontier cross-disciplines: information science theory, system science, control science, cognitive science.

Basic level of specialties: electronics, optoelectronics, microelectronics, digital technology and computer science, information and signal processing science and technology, and bioinformatics.

Discipline level: a variety of subject areas, such as communications, remote sensing, microelectronics design, electronic technology, and design and manufacture of optoelectronic devices.

4.4.2 Social Supports and Major Institutions Undertaking Tasks Necessary for the Development of Different Levels

Basic level: support from the National Basic Research Program and special state funds; tasks mainly undertaken by universities and basic research organizations.

Basic application level: 10-15 years of medium-term and long-term pre-researches; 5-year pre-researches are mainly supported by the relevant national plans and funds with supplemented supports from large enterprises; tasks mainly undertaken by the institutes as that in the basic level. Application research and development are mainly supported by enterprises with supplemented support from the state; tasks are mainly undertaken by enterprise-based R & D institutions.

Service application level: mainly supported by enterprises and enterprise-based R & D institutions (market applications and support, security support, etc.)

4.5 Examples of Typical Information Systems and Their Main Characteristics

4.5.1 National Defense Information Systems

The role and significance of information science and technology and information systems in the field of national defense and military can be illustrated in the following aspects. Tracing back to 2500 years ago, in the Warring States Period, Sunzi already pointed out in the *Attacking by Stratagem* in *The Art of War* that "know your enemy and know yourself and you can fight a hundred battles without peril". This embodies the basic law of war followed by a man from ancient times to the present (including the modern information war) in the fields of military theories and philosophy and contains information as well. "Know your enemy" is achieved by access to "information", but the accesses to information and equipment in ancient and modern times are different—with the support of modern information science and technology, military equipment is informationized. Informationization of equipment means embedding information technology and information subsystems in weapons and equipment to improve performance and forming more advanced and more complex large-scale weapon systems by integrating information systems (informationized equipment) with other weapons—it can also be called equipment informationization. Informationized equipment refer to military information systems mainly formed by information science and technology which has certain combat effectiveness, such as military reconnaissance satellites, electronic warfare systems, and so on. One important symbol of military modernization is a high degree of equipment informationization and a large number of informationized equipment. The essence lies in the acquisition of the enemy's situation, and timely, effective measures to combat and strike.

In informationized warfare, the higher degree of informationzied equipment and equipment informationization requires more attention to information security and confrontation. The war itself is a fierce confrontation, so "confrontation" is the essential element of the war. Informationized equipment and equipment embedded with large-scale information science and technology encounter the problem of working in confrontational situations. What requires more attention is the double-edged sword effect—the opposite might make use of the characteristics of "information" in equipment to serve themselves, thus resulting in damage to our side. This relates to the main topic of the book and will be discussed in later chapters.

4.5.2 Recent Developments in Information Systems for Civilian Use

Home (individual) information systems: there is a trend of information equipment for entertainment and everyday life integrating with office information systems.

TV Terminal: televisions connected with television networks (broadcast and cable networks) are transforming from the first generation of analog signal to digital signal; digital television (with clarity of about 800 lines) will promote at a faster speed than high-definition format television (HDTV), while television, as the video system of terminal equipment, will be further developed, for example, forms "home theaters" with a new generation of DVD, or video systems that better meet the individual needs through combination with National Broadcast & TV General Bureau's radio and television on-demand network.

Computers and computer networks (including the modified versions): further development of a variety of services (including life services, business services, and information services); exclusive advantages of electronic information, computer networks, radio, and television networks. In terms of infrastructure, telecommunication networks and computer networks have shared areas, but applications are still clearly separated; how to share resources for the development of more kinds of services and improvements in service quality remains a daunting task.

Phone and personal mobile phones. The mobile phone business has increased rapidly. The number of China Mobile subscribers ranks first in the world. The main service is the voice communication service, with others as supplementary services. Individual broadband demand will gradually increase, so the main task of 3G and 4G mobile systems is to ensure high-quality voice communication service for a large number of users and provide multi-service communication systems. Mobile phones are small and beautiful and have ornament features (especially for young women). The number of fixed phone users increases with the continued growth in service types, although the growth rate decreases.

Integrated Intelligent Domestic Services System will further develop. One possible way of developing high-end residences services would be a smart integrated service system network within which the domestic electronic devices are connected through a wireless mini network. Wireless mini network IEEE80211BWLAN, "Bluetooth" and another short-range communication system will be integrated to form a comprehensive service system with better function.

China has a large population and numerous families. Such a situation offers an extensive potential market for the above-mentioned information systems for family or personal use which will form an industry with large output and is of great significance for economic development in China. Besides its importance, some problems are worthy of attention.

Development projects which enable wider bandwidth and better effects, such as the convergence of 3 Networks, and Fiber-to-the-Home/Building, will be promoted gradually.

Centralized Chinese residences, especially the new residential quarters, have better conditions and thus are more possible to be equipped with more advanced information systems. Also, because of the centralization of users, the implementation of broadband in residential quarters or apartment buildings is easier with lower costs.

Some lines are poorly constructed and have quality problems, such as strong standing wave, large contact resistance, and serious cross-modulation crosstalk—these affect the quality of high-speed information services and thus require part rework, and increase spending.

Overall, the use of information equipment needs to further reduce costs (installation and operating costs) and promote the use by social development.

High-end core products and security products in the computing field. These products should carry out independent research to develop safety products with intellectual property and high reliability, to form an industry that ensures the application of critical information equipment and network security (including software, hardware, chips, etc.)

Critical equipment of special information equipment and corresponding software can form application software development parks and software industries, and development in this field should be speeded up.

The third-generation mobile communication has key chips and systems with independent intellectual property and will be developed accordingly. With the increasing scale in opening up and advance of the economic development model, as well as the industrial upgrading and fierce competition in the international export market, information systems become essential for quality improvement, cost reduction, services improvement, and management improvement. Against such a backdrop, Chinese enterprises alone have a rather great demand for information system development.

A variety of microelectronics chips (including SOC, etc.) for other purposes (apart from computers), and various special embedded systems, such as transport cards, chips used in a variety of electrical appliances, automotive electronics, and information products.

4.5.3 Discussions on Some Basic Concepts of High-Tech Enterprises in the Field of Information

Developments in the field of information hi-technology are rapid and provide very attractive benefits. But it should be noted that rapid development indicates, on the one hand, much room for development in recent demands; on the other hand, immaturity, instability, uncertainty, and rapid change in performance, norms, standards, unstable demand, and a number of users contribute to the negative factor of "market risks". Enterprises in this field should have strong comprehensive strength and correct entry points in the market to be successful.

High efficiency attracts investment, manufacturing, management, service, and heavy involvement in relevant areas. On the one hand, it forms strong competition; on the other hand, it is also very easy to cause oversupply in the market which would eventually lead to reduction and adjustment. In the economy system, excessive development and growth of the information industry will lead to a bubble economy, and result in an imbalance among the economic sectors. In the case of long-term imbalance, the bubble will burst and affect the normal development of the economy.

For the basic facilities, such as power, transportation, and core telecommunication facilities, larger room for development should be left while meeting the current needs. For the application and consumer products, development should keep pace with the actual demand; the products for the ordinary consumers in special, must not be developed according to the sophistication of technology—the developers should take full account of the consumer demand and consumption level. For example, the "Iridium" system for mobile communications has advanced technology, but the cost is unaffordable for ordinary users. This is exactly the reason for its fast and miserable failure.

The one-sided, extreme and exaggerated propaganda made by market advocates, experts in the field, advertisements, or even some well-known enterprises and famous people, should not be believed readily. Absolute assertions are unrealistic and superstitions should be avoided. Science analysis should be upheld. Examples of such kinds are numerous. Seven or eight years ago, a blare of publicity said that "Information highway and multimedia applications will be soon realized and B-ISDN will replace all other systems." But today, the U.S. telephone network is still developing at a certain speed (XDSL and other related systems are not B-ISDN systems), and although planned multimedia cable network is under development, there remain many problems to be solved and it would take long before it completely replaces the telephone network and cable television network. The key reason is that the demand for broadband applications with a bandwidth of several or dozens Mpbs among American households is not strong. The advertisements said that ATM would replace the other exchange systems, that broadband multimedia service for 3G mobile communication users was imminent, and that the high-speed INTEL network would become the mainstream of communication services, but they are all proved wrong. In 2000, propaganda claimed that "Bluetooth" technology would sweep the world in the summer of 2001. But that was a far cry from reality. Examples of this kind are just too many! They prove the developments of the information systems, but that these developments are not simply determined by certain technical merits or advantages—they must keep pace with the development of human society.

4.6 Conclusion of this Chapter

This chapter focuses on basic concepts of information, information science and technology, information systems and related contents, including: the definition, connotations and development process of information; the characteristics, functional components, ultimate goals of information systems and the large and strong supporting systems needed for their development. Some typical examples of information systems and description of some basic concepts about the development of high-tech enterprises are given in the end.

Exercises:

How do you define "information"? What are its connotations?

What are the characteristics of information? What is the history of human using information?

What is information science? And information technology?

What are the basic features of information systems? What is the ultimate goal for the development of information systems?

What is the support system for information system development?

Why is information technology and information systems development one of the eternal themes of human society?

What is the general function of information systems?

Explain the process and function of information collection, transfer, processing, exchange, storage, control and management of a certain information system, and its development process and support system.

Analyze and demonstrate the characteristics and functions of embedded information system?

Chapter 5
Basics of System Knowledge, Representation and Description

5.1 Introduction

"System" is a combination of things in complex motions in which things have multiple levels, are in complex dynamic states and mutually intertwined, and display dynamic characteristics in general. "System" is objectively existent in human society, in particular, the continuous generation of "artificial system" is of significance to human development. In order to master and develop the "artificial system", we must have rich knowledge of the "system", i. e., get a comprehensive command of the important principles of its evolving from the general to the inner on different levels and sections. The basic principles of "system" naturally refer to the system theories in development, including the theory of dissipative self-organization theory, etc. (which had been discussed in previous chapters). This chapter discusses from the perspective of application some important concepts, methods necessary for the knowledge of systems on the application and application base levels, and the essential mechanisms in artificial systems as well. This chapter will combine all of these to formulate system representation and description.

5.2 Basic Concepts of System Description and Representation

5.2.1 Proper Understanding of "Human-Centered" Concept

Human beings keep deepening their understanding of both objective laws and humans themselves via thinking, then putting these understandings into practice to create new things. In the world of human harmonious development, "human-centered" is the most essential concept in which "thinking" is the core element. Although basic rules existing in the motion of objects are independent of human thinking, their uses can not be brought into full play without being understood and applied by human beings. Meanwhile, the connotations of rules, especially the "hidden ones", rely on the development of human society to become from covert to overt existence, and thus explored and understood by human beings. Therefore, "thinking" is the process and phase achievement in the constant development of human initiatives based on objective laws. It is also the core element displaying human abilities. The artificial system is an important manifestation of the "human-centered" concept in the process of human development. And the development and proper application of "thinking" is the in-depth manifestation of this concept.

5.2.2 The Concept of "Relations"

In "system", an existing thing, exist "motions" different from others. And in motion, there certainly exist interactions and mutual influences with other things and eventually formulate phase results. These are all called "relations" of things. Some of the essential relations or relation sets can represent the motion of things' survival, the laws, and the dynamic process of motion. Therefore, the basic concepts in the representation of the system are dependent on its key "relations". Now system theory maintains that the relation sets having self-organization mechanisms with the system's dissipative structures mutually interacting with the environment are called self-organization relations, which are an important part of the system representation. "Relations" feature a wide range of varieties, experiencing the existence of things and bound to have their distinctiveness. The ancient Chinese philosopher divided relations into two categories: the "tangible" and the "intangible", to embody the complex temporal and spatial relations in the motion of objects and their dialectical development. The "intangible" means there is no fixed existence, indicating the future state and various complex possibilities in time-space dimensions, such as the relations represented by "capability", and by "hidden dangers". These kinds of intangible relations are of great significance in studying the survival, development, and changes of the system. They certainly deserve our careful studies. In addition, equivalence relations and similarity relations are also important. Equivalence relations signify a generalized similarity in essence, and can be used as a gauze to making classification; similarity relations are more reduced than equivalence, which does not have the same transfer characteristics as equivalence relations do (i.e., a~b, a~c, then b must be c, ~ means equivalence relation).

Various transformation relations, constraints, and symmetry relations are also important relations.

5.2.3 The Concept of "Mapping"

"Mapping" is a concept to manifest the corresponding relevance in different areas. It borrows from mathematics. In mathematics, mapping and the concept of "set", constitute the most basic concepts of modern mathematics. The mappings of various properties, such as surjection, injection, one-to-one mapping, and homomorphic mapping, represent the important relations between two sets and the internal relations within two sets, and can also represent the interrelationship and interaction between and within different components. For human beings, the essential element in the process of learning is to understand and make use of "concept", which is formed in the process when the motion of things turns into a mapping in the brain. To form a correct "single mapping" is an essential issue. For man, before the cognition system turns into development system, description system is necessary, which is a descriptive system to make profound research into "system" and its changes.

5.2.4 The Concept of Operation

The interrelations between and within systems can be transformed and represented by mathematical (algebraic) operations, with which the process and result of system's related interactions and results can also be represented by mathematic operation. The mathematic expression of "operation" is $A \times B \rightarrow D$, in which A、B、D represent three sets, and $A \times B$ represents the interrelation between the elements within A and B, which occurs according to certain rules; the result is the mapping (represented by $\rightarrow$) to D set. "Operation" is characterized with universality and stringency, and in different situations "$\times$"and"$\rightarrow$" refer to specific rules in order to indicate the specific interrelations and the corresponding results.

5.2.5 The Concept of Generalization, Extraction, and Abstract Modeling

The key concepts used to characterize the description system can be concluded as generalizing all the important clues, and situations (no important situations are missing), extracting the important elements which cannot be ignored, and then abstracting them into important relations, on the basis of which an appropriate correct model is to be established. "Appropriateness" means no redundancy and having no unnecessary contents, and "correctness" means no missing information essential to represent the description system. Building an appropriate and correct system is the foundation for the following-up research system. What's more, the whole system, which is under the guidance of generalization, extraction, and abstract modeling, requires repeated examinations and revisions on a multilayer in order to give full scope to its functions.

5.3 Preliminary Study on Representation and Description of System

5.3.1 General Overview of System Representation and Description Fundamentals

The overall characterization of the system is mainly based on the core mechanism and the corresponding structural composition of the overall survival movement of the system, in addition, the expression of the survival conditions should not be omitted, so the overall characterization of the system should be based on the open environment, the survival conditions based on the structural model of the system layer self-organizing mechanism composition. The theoretical basis for this characterization and description of the system is the modern dissipative self-organization theory. Since the system is in dynamic motion, the environment and conditions are changing dynamically, therefore, combined with the practical application of the characterization system, the self-organization mechanism should be extended to the active self-organization mechanism for characterization and description. Since the system is now mostly a spatio-temporal multi-level and multi-profile dynamic structure, it is often necessary to characterize the system in detail with multi-level and multi-profiles in combination with practical needs, and also to use the dynamic self-organization mechanism embedded in the multi-level Spatio-temporal domain expansion model of the system motion as described in Chapter 2, Section 5 of this book to characterize the system in a hierarchical and specific way according to practical needs.

5.3.2 Basic Thinking Modes in System Representation and Description

5.3.2.1 The Basic Mode of Thinking—the Combination of Qualitative and Quantitative Methods

Qualitative and quantitative thinking are two basic modes of thinking (forms). The combination of both constitutes the basic thinking of human beings. A lack of either or an improper combination can not make scientific thinking. Its importance is due to the fact that the motion of complex things (system motion) is closely related to each other on multiple levels and sections. I. e., profound quantitative researches and comprehensive studies on the general are necessary in order to grasp the rules and features of the overall motions. The combination of quantitative detail studies and studies on the general is indispensable to a scientific command of the complex motion of different things. After learning its significance on the theoretical scale, what matters next is to learn how to integrate them together in practice, yet at present, it's difficult to crystallize how to integrate them together. Nothing but repeated verifications and examinations of the results in concrete practice in the process of representation can achieve the target. Besides, qualitative and quantitative thinking is usually integrated together.

5.3.2.2 Another Basic Mode—The Combination of Analysis and Synthesis

"Analysis" is a general name referring to the way of thinking modes starting from the whole to penetrate individuals in the study of certainly associated movement mechanisms. From the perspective of movement mechanism, it is necessary and natural to do analysis, but the problem is that matters are universally connected with each other and that the effect on different issues and associations might be different. Therefore, finding out the main relations and then doing an analysis is the prerequisite for making correct conclusions. And the extension of the movement and complex factors may lead to overstressing of "reductionism ", and become less scientific but subjective. "Synthesis" refers to the thinking mode that starts from individuals and induces a conclusion that is applicable to reveal the whole tendency and mechanism. Therefore, synthesis is also an important way of thinking, but its defect is that it tries to cover every aspect of a matter so that the key point can not be elaborated on or given enough account. Taking the above into consideration, the two methods are opposite to each other, but also complementary, which is a scientific combination. How much scientific it is depends on the concrete combination of its elements, that is, to what degree the elements contribute to the whole.

5.3.3 Brief to Some Commonly Used Basic Methods

(1) Build the general model and model combination of system representation and description using relations mapping and inversion under the guidance of thinking modes featuring the combination of qualitative and quantitative methods, and that of analysis and synthesis. See Figure 5-1.

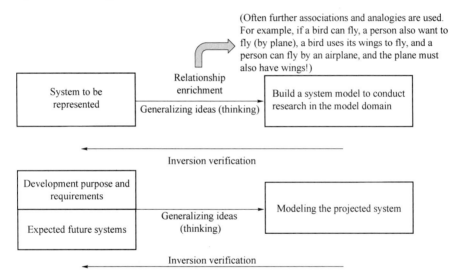

Figure 5-1 Diagram of the method of inversion through relation mappings

The established model can be applied to induction and the development of new systems. The key to the success of "inversion modeling through relation mapping" lies in selecting the essential relations to reflect the modeling, and then carrying research in the domain of models, i.e., the research result drawn from relation deduction is fed back to and verified repeatedly in a practical physics domain, and then models representing a certain type of system according to concrete requirements can be established after repeated verification. If the model is to represent a future system, it should be inverted to physics region for practical operation after the operation and arrangement in the model domain. And the general scheme is decided after repeated feedback. The method of inversion through relation mappings is a basic method, and more about it will be discussed in chapter 7.

(2) Build models sets according to sections of corresponding layers respectively and make them as complement for the general model, represent and describe system in detail based on different layers and sections to form the model set, and make further complementary representation to system, i.e., supplement the effects of key factors under the guidance of general model, amendment to the general model is usually made. See Figure 5-2.

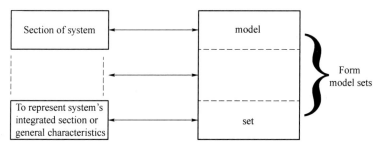

Figure 5-2 **Hierarchical, sub-section modeling mode**

(3) Achieve representation and description using multi-types of models on the basis of modeling types

The key types of models are as follows:

➢ Mathematical model based on the representation of a mathematical method
➢ Model integrating mathematics and physics
➢ Model integrating physical simulation and the part in a kind
➢ New system based on the integration of original system and the Improvement model
➢ Scale model in scale condition representing system

The functions of the model can fall into several types, such as the black box model without considering the internal structure but based on the input and output characteristics in the representation system, the model of the behavior emphasizing structure and state of the system, etc.

With the application and development of the model, besides using differential equations to represent the continuous dynamic behavior of the system, the computer is applied to build the space (state)-time-double-discrete input and output—system state transfer six tuple model:

$$S_i \longrightarrow \boxed{F^n(S_i, Q, \delta, \sigma, C, S_o)} \xrightarrow{S_o}$$
$$C \nearrow$$

Where S_i, S_o is the input and output vector, or quantitative sequence, C is environment sequence, and Q is the system state sequence, δ represents the system state transfer formula as shown below:

$$\delta: Q_{i-1} \times C_{i-1} \times S_{i-1} \to Q_i$$

$$\sigma: S_{i-1} \times Q_{i-1} \times C_{i-1} \to S_{oi}$$

> One must pay close attention to whether the dynamic processes of the system is "correct" and "appropriate" after building the models, it is the core to determine whether the model is "correct" and "appropriate" or not. Besides, there exist different types of model composition for the combination of models, but their description of the process of dynamics must be linked up without contradictions.

(4) Describe the principal characteristics of the system to be described via borrowing from the existing systems. —Notice: Attention should be paid to the similarities and differences of constraint conditions and intrinsic mechanisms between the two systems.

Case 1: The representation of the Chinese express railway, the world's longest express passenger railway, its velocity of 300 km/hour can compete with some countries but its operation distance is up to thousands of kilometers. Its key characteristics are long distance and high velocity in the whole journey, which increases the difficulty of the system implementation. So more requirements and more constraint conditions for the description and representation of the express railway according to the practical conditions are needed, with which long-distance express railways anew successfully built.

Case 2: Large reservoirs with water retention capacity reaching billions cubic meters, generated energy reaching billions kilowatt-hour and able to resist once-a-century flood. If designs of such reservoirs are based on the enlarged size of certain prototype without considering the high composition of silt in the water flow of these reservoirs, fundamental mistakes will be resulted in on the description and representation of system's key characteristics. Consequently general schemes have to be altered and heavy losses are made.

5.3.4 New Field Methods in System Design, Representation and Studies on Complex Scientific Problems and Development of Technical Innovations Based on the Guidance and Support of Computation Science

The method described in this section is a research methods system based on the thinking mode combining qualitative and quantitative methods, and the analysis with synthesis, and supported by the relations inversion and mapping in the basic disciplines like mathematics, physics, computing science, computer science and technology and control science and technology, and the technical sciences, etc. In 2005, President's Information Technology Advisory Committee (PITAC) submitted to the president to name it "computation science", and regarded it as the core requirement for American competitiveness. Great importance was attached to it.

(1) Computation science as the key technology method.

Computation science indicates solving problems in the real world via computation capability. The real world is intricate, in which some aspects are very complicated to man's knowledge (They are objectively existent, including being-in-itself for man, but far from self-understanding.), which cannot be solved merely depending on theoretical research and experimental verifications. Meanwhile, the above "complexity" refers to not only the complexity of the motion laws of the studied subjects, but also the complexity formed in environment and condition domains, e. g., the complexity of generating conditions, the extremity of environmental conditions, the arduousness to realize the conditions for experiment and to get the results of the experiment, and etc. This invalidates the thinking and working modes that results can be achieved via repeated experiments. In virtue of this, man gradually found and eventually proposed the third way, i. e., solving complex problems depending on computation science. Computation science (not computer science) is a disciplinary domain whose connotations include multiple disciplinary levels, having a wide application range and comprehensive application characteristics, e.g., solving complicated scientific problems (e.g., biology, physics, chemistry, climate, and meteorology), engineering problems (e.g., large-scale artificial systems' research and development, large-scale civil aircraft, dams, nuclear power plants, aviation engineering, and national defense systems), researches on social problems and humanities (e. g., non-traditional security problems, such as diseases, natural disasters, social economy, problems of wealth and poverty and population), and is an "indispensable" method.

(2) Connotations of computation science.

Computation science comprises the following three parts:

> Algorithms (including numerical and non-numerical), modeling and simulation software necessary for the solution of problems in science, technology and humanities science

> Developing and providing accented supports to computer science and information science; Optimizing the hardware, software, networks and computation in advanced systems, and data relevance

> Development of infrastructure for computation science (including high-level professions)

Chapter 5 Basics of System Knowledge, Representation and Description

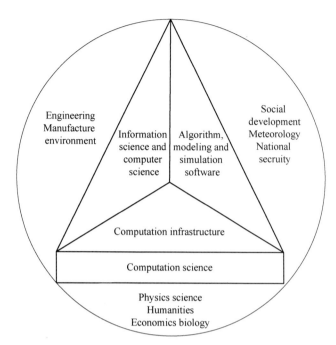

Figure 5-3 Connotations of "computation science" and the related connotations
(stated in the report by PITAC)

(3) Characteristics of Computation science used in system representation and description.

Computation science, as the third field (besides theoretical research and experiments), is actually the process-based and scientific representation and description of complex systems and systematic problems with "computation science" as the core (not including all matters, big or small). Here the "process-based" includes a combination of analysis of representation and synthesis of application, a combination of repeated modeling and feedbacks and verifications, combination of parts and whole, a combination of multiply distributed parameters and important sequence parameters as well as the combination of historical accumulation and current state (continuation of distribution in the time dimension), etc.

The kernel in describing the complex system is relations mapping modeling. Present models involve the combined modeling of multiple fields and disciplines (Not that experts in certain individual disciplines construct models.) Models need to reflect the motions in forward and reverse directions in the process of motion, which is of great importance. Specifically, in modeling multiple parameters mold the structure of the model, parameters of variation verify the model, and the verification of models optimizing the application of models is also of great importance concerning many aspects. Comparing (or partially comparing) the model simulation with the practical results is one of the key methods used to verify models.

The solution of complex models depends on the algorithm and optimization of the algorithm, and the comparison and integral application combined with the database (using the data in the database, and the simulation reproducing history.) As a result, the algorithm is the second most important integrant after modeling (Sometimes, the algorithm can be correlated to the structure and application characteristics of computation appliances.)

The detailed solution process of complex models and multiple parameters needs high-performance computation appliances as computation infrastructures, including hardware, software, and the fine computation structure combining hardware and software. And high-level talents of various important disciples who adapt to the application development of computer science are also included. In light of the above, system representation and description and application supported by computation science are the extensions in the time dimension of the above-discussed thinking mode, methods and basic methods, and the development and application via high-performance computation infrastructures.

5.3.5 The Extension in Applications of System Representation and Description

System representation and description are the basic integrants in studying systems and are tightly related to further applications. Actually, all the above sections are concerned with applications, particularly section 5.3.4 in which systematic methods able to be supported by computation science are discussed. The extended applications of system representation and description can be summarized as follows:
- ➢ Extension and performance optimization in applications after the analyzing and representing the current systems
- ➢ Paint a blue print for the general planning and structural composition of the research and development of system based on requirements for functions and constraint conditions before developing new systems, and analyze and synthesis multi-round research and development process
- ➢ Advanced and detailed management of the operation of the system with system

representation and analysis as the necessary basis; e.g., to guarantee security when the system is running, detail analysis of the fundamental structure and possible security loopholes is essential.

5.4 The Theory and Method of Living Self-Organization Mechanism, and the "Multi-Living Agent"

5.4.1 Brief Introduction

The proposition of multi-living agent theory and method aims at extending the dissipative self-organization theory on the basic level into application bases levels, then forming the comparatively practical theory of understanding systems, and being further used to represent systems. The ultimate goal is to propose a new method to study and design new artificial systems (to be discussed in chapter 7). The key points are briefly illustrated as follows:

(1) Extend the dissipative self-organization theory, starting from the core characteristics of "system's" existence, and regard system's self-organization mechanism as a thing based on the principle that the existence of all things is relative and conditional. Therefore, the existence of this mechanism is also conditional. And in order to cope with the existing problems of system's self-organization mechanism in the actual environment, dissipative self-organization theory develops into living self-organization mechanism and its representation is given.

(2) According to the principle that the method of analysis and synthesis are equally important to the acquisition of knowledge on complex things, we should decompose the characteristics and features of a complicated system into those of several sub-systems; and integrate the dispassive characteristics into the characteristics and features of the complex system as a whole. In order to "decompose", "synthesize" and represent the complex system, the author developed the method of multi-agent in the computer field into "multi-living agent method and theory". The basis of system livelihood is actually the living self-organization mechanism, which can be decomposed into the livelihood of its agents. The consultation on three levels (including the man-machine cooperation with administrator's involvement) can represent the comprehensive function of multi-agent system. "Agent" here connotates the concepts of human intention, some system functionsand system structure with the connotations of

"livelihood" and forms a "representation" of parts of system (Please refer to 5.4.2 for the connotations of "representation".)

(3) The research on the function of livelihood of agent functions does not focus on the instantaneous maintenance of function, but on extending the functioning time of the system in the environment which affects the system functions or system survival (including the constant change of environment). When extended to the field of design, the method of multi-living agent design is not a status design that satisfies a system with given indexes, but a process design that satisfies given indexes in a certain process. This is an advanced design idea.

(4) To realize process design, apart from taking the measures pre-determined to improve functioning in the design stage, we should regulate certain important systems in terms of the environment, the state of cognition, and the livelihood function (in order to formulate an extended livelihood function mechanism), to further improve the livelihood of the function of the system and apply this function in practice to mediate in and among agents. In concrete analysis, we should fully employ the method of computer science mentioned in 5.3.4.

5.4.2 "Livelihood" and Agent Livelihood

(1) "Livelihood" in Chinese refers to the characteristic of the life characteristics of living beings. It has a wide range of meanings and possesses image characteristics. Its antagonistic term is "livelihood loss" (indicating "death"). As for humans, there is a variety of deaths with a lack of 'livelihood', such as physical death (marked with the stop of heartbeats), brain death (loss of cognitive function), the death of thought (Thoughts are extremely backward or decadent; for example, a small number of Japanese extreme right-wing politicians are still obsessed with Japanese military aggression and its oppressive invasion into East Asia.), the death of ethics and morality (such as murdering or harming others against morality for personal interest), etc. The more complex the living body is, the more complicated the characteristics of its "death" is. The main parts of artificial systems are made by human beings, i.e., artificial systems are not living beings, therefore, they have no life characteristics. But some important artificial systems are closely associated with the existence and life of human beings, so they are of great significance. And so, the study of artificial systems is rich in significance.

(2) Analysis of system's livelihood from the functional level.

The characteristics of "livelihood" are divided as normal-weakened-approaching zero-the emergence of forbidden negative functions

"Livelihood" is divided into the following levels:

①Basic factor level.

②The level of necessary and sufficient conditions, and conditions correlating with others.

③Direct function level.

④Agent survival level.

The scope is divided into: general, partial, and local.

The above function livelihood characteristics can have further division, which proves that subtle researches on the function livelihood characteristics are complex. Now further discussions involving "agent" and "multi-agent" will be made. "Agent" derives from the field of computer software design, and the "living agent" means to integrate human intentions and the structure of systems on the living level, and finally achieve certain functional purposes. When a "living agent" in an artificial system "exists" in the operating method of functioning, it is mainly composed of the "tangible" form of the structure which is constructed by hardware and software, and the "intangible existence" which is composed of man's related concepts and the mutual "relations" they formed. Under this condition, "tangible existence" is important. But during the formation process of a "living agent", its existence is named by the author "model register", that is, the model calculation that combines computation science and computational plane based on thought model (application of relation-mapping-inversion method). In the simulation environment, "multi-living agents" are combined with several agents with livelihood; they function respectively through mutual consultation and coordination and thus form the function livelihood of the whole artificial system.

(3) Types of living agents and livelihood loss.

The basic concept set by "agent" is wrong or its vitality is severely impaired; on the basic level the positive approaching zero or negative existence, which equals to thought death of human.

The loss of preconditions of agent function (including correlating with other agent conditions): The condition level approaching zero, which is tantamount to the vegetative state.

Loss of agent function or "agent" turns into an inexistence state: function or existence state approaching zero, which is tantamount to human death state.

The agent function turns to the reverse, and its characteristics are its property turning to be negative, which emerges in the direct function level. The aforesaid types of agent livelihood loss can exist on the system level which is composed of multi-living agents and in the range of an individual living agent as well. Therefore, the loss of livelihood can be on a partial scope or an overall scope.

5.4.3 Expressions of "Agent" in Dynamics (Function State, and Environmental Constraints, etc.)

The dynamics expressions are a set of multi-variable nonlinear differential equations which describe the "existence" and "operation" of things. It is rather difficult to solve the equations. Here, we mainly discuss the basic concepts and adopt the functional analysis to express the dynamics of the agent as follows:

$$LA_t(Q_t, S_{it}, \delta_t, \sigma_t, S_{ot}, C_t)$$

where Q_t represents state of LA_t at time t; $S_{it} = S_{it}(S_N \times S_A)$ is the input flow at time t, S_N is the normal input control flow, and S_A is the in-put attack flow that can infiltrate into the normal flow to obtain the aim of attack; S_o represents the output flow of LA_t; $C_t = C_t(C_A \times C_N)$ is the flow of environment information of LA_t at time t, and it includes the normal control parameter of LA_t, the support condition flow C_N, and the interferential control flow C_A of the attacker; δ is a mapping relation from the parameter of LA_t to the state of LA_t; which can be used to represent the livelihood of the agent at different times and can be described as follows:

$$\delta: Q_t \times S_i \times C \rightarrow Q_{t+1}$$

σ is a mapping relation from the parameter of LA_t to the output flow of LA_t. It is used to describe the changes of output flow, and can be described as follows:

$$\sigma: Q_t \times S_t \times C \rightarrow S_o$$

When dealing with concrete issues, the above relations (including principles in selecting specific elements from the Descartes set, specified forms and principles of the mappings, and the relations mentioned in the following chapters), the operation within the set, and the identification of mappings are really difficult and complicated. These problems are among the complicated issues that scientists and technology experts focus on and are under arduous research yet still have no fundamental breakthroughs.

Therefore, no detailed discussion is possible here but a general description (The work done by the research group with this author will be introduced in the future for help and comments.). For the multi-living agent, with the influence of its initial condition and environment information flow, the dynamic evolvement of the input information flow through mapping is possible. The dynamic evolvement of the function of the living agent can be illustrated as shown in Figure 5-4.

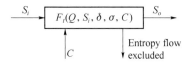

Figure 5-4 The dynamic representation of living agent

5.4.4 The Living Self-Organization Mechanism (LSOM) and the Expression of Two Set Models

(1) Self-organization Mechanism.

We believe that the sectional characteristics of a system at different times can be represented by its relative corresponding self-organization mechanism, which in mathematics can be expressed as follows:

$$R'_I(r_{I1},\dots r_{IS}, t, T) = R_I(R_1 \times \dots \times R_n, t, T_1)$$
$$R'_{II}(r_{II1},\dots r_{IIS}, t, T) = R_{II}(R_1 \times \dots \times R_n, t, T_2)$$
$$\vdots$$
$$R'_M(r_{M1},\dots r_{MS}, t, T) = R_M(R_1 \times \dots \times R_n, t, T_M)$$

Where R'_I is the self-organization mechanism in profile I, and is made up of $r_{I1},\dots r_{IS}$. In this expression, t represents time variant, T reference time, they are composed of the Descartes operation results of the relation elements $R_1\dots R_n$; use "agents" to represent the "system", then the above expression can represent the self-organization mechanism of "agents".

For example, the above relation group can be expressed in dynamic six elements relations expressions:

Self-organization mechanism of the output profile: $S_o = S_i \cdot \sigma(S_i \times Q_i \times C)$

Self-organization mechanism of the state profile: $Q_{i+1} = Q_i \cdot \delta(Q_i \times S_i \times C)$

In the countermeasures environment:

$$S_i = S_i(S_{iN}, S_{iA})$$

$$C_i = C_i(C_{iN}, C_{iA})$$

The normal self-organization mechanism of the agent can be represented as follows:

$$S_{oN} = S_i \cdot \sigma(S_i \times Q_i \times C_i) = S_{iN} \cdot \sigma(S_{iN} \times Q_i \times C_{iN})$$

That is to say, to maintain the normal self-organization mechanism in a countermeasures environment, the prerequisite is as follows: try to make the countermeasures factors not work or infiltrate them, that is, $S_i \times Q_i \times C_i = S_{iN} \times Q_i \times C_{iN}$

Another example: system (agent) output abiding by the slaving principle given by Professor Harken.

$$S_{ii} \longrightarrow \boxed{LA_i(Q, \delta, \sigma, C, S_i, S_o)} \longrightarrow S_{oi}$$

If $S_i = F(t)$ $LA: \dfrac{dS_o}{dt} = -rS_o + F(t)$

Inferred from $F(t) = ae^{-\xi t}$, the stability of $S_o(t)$ can be expressed as follows

$$S_o(t) = \frac{a}{r-\xi}(e^{-\xi t} - e^{-rt})$$

If $r \gg \xi$, then $S_o(t) = \dfrac{a}{r}e^{-\xi t} = \dfrac{F(t)}{r}$. It proves that under certain condition "input" can make "output" follow it ("input" enslaves "output" to follow it), then $r \gg \xi$.

(2) The living Self-organization Mechanism (two set models).

According to the above section, the self-organization mechanism is composed of conditions, and the "existence" of all things is a unity of opposites, in which "unity" signifies that one side suppresses another to form a unity of the dominating sides (The properties of things are determined by the dominating side.) For further analysis, the rationality of the dominating sides results from the rationality of the sufficient and necessary conditions supporting them, and meanwhile, the necessary and sufficient conditions for the opposite sides are necessarily suppressed and can not give full place to their functions. As a result, the research on living agent self-organization mechanism (LASOM) can be evolved into the study on how to construct sufficient and necessary conditions for LASOM or the study on how to suppress the conditions affecting livelihood mechanism. Therefore, we establish a two-set model to represent the LASOM, which is composed of condition set A and self-organization mechanism set B. The operation and mapping of the elements or subsets of set A to set B form a self-organization mechanism and represent its livelihood as shown in Figure 5-5.

Chapter 5 Basics of System Knowledge, Representation and Description

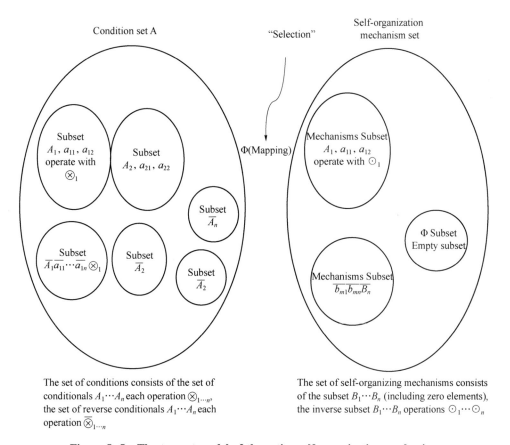

Figure 5-5 **The two-set model of the active self-organization mechanism**

Assume that the self-organization mechanism is composed of some sub-mechanisms and components which are represented by subsets and elements. Mechanism sets also include some empty subsets, and reverse set $\overline{B}_1 \dots \overline{B}_n$, which is used to represent that the self-organization mechanism is turning negative and lost. They work together to represent the existence and loss of the livelihood of SOM, a step further, they correlate with the existence and changes of SOM depending on corresponding conditions. Actually, the function of "condition" is rather complicated, so it is classified into three types: positive, negative, and zero. The specific operation is controlled by various operations and mappings, such as $(a_{11} \otimes_1 a_{1n}) \xrightarrow{\Phi} b_{mn}$ or $\overline{b}_{mn}$, $a_{11} \overline{\otimes}_1 a_{1n} \longrightarrow$, or $a_{mn} \xrightarrow{\Phi} b_{mn}$ or $\overline{b}_{mn}$. What's more, in model application, attention should be given to the issues of the selection of conditions, subsets and mappings, which is the practical expression that the interaction of multiple factors affects the livelihood of SOM. Further research and analysis can be carried out with clear concepts after the establishment of the "two-set model".

5.4.5 The Composition of Multi-Living Agent of System and the Three-Set Model

(1) Three-set model of agent representation system.

On the basis of two-set model representing SOL, we establish a three-level (composed of three sets) three-set model in which agents and their conditions constitute MLA representation system. It's illustrated as shown in Figure 5-6.

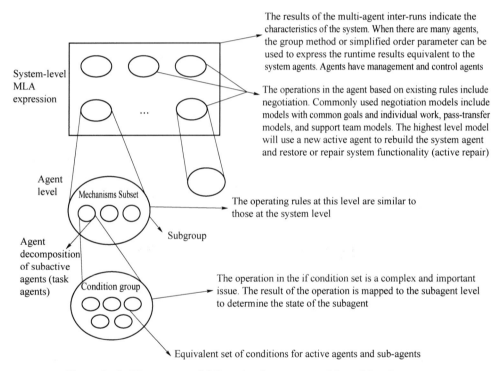

Figure 5-6 Three set model for complex systems with multi-active agents

(2) MLA representation system structure.

The above structure is universal and reducible in which the multiple agents constitute the basic level, the middle level is the input and output interface layer, the higher level is the management layer hidden behind the background which can be further divided into two layers: One is the negotiation and coordination between multiple agents in system layer generated by the management and control agent (The normal mode is that laws of logics and the expert systems co-support the coordination of the agent.); the other is the administrator decision scheme which centers around the administrator, with management and control agents being the assistance (such as providing information and options for the decision scheme). The function of adjustment is maintained on the basis that each agent has functional living self-organization mechanism based on information and laws of logics. Therefore, the artificial system has coordination of four levels of functional living SOM and can eventually serve its purposes. See Figure 5-7.

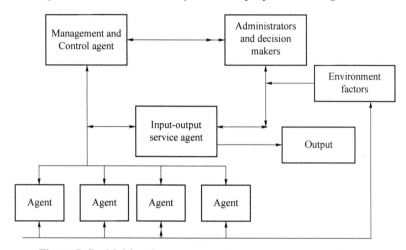

Figure 5-7 Multi-active agent complex system structure diagram

5.4.6 Quantitative Expression of the Livelihood of "Agent"

(A) Using the sufficient and necessary condition for the existence of agent function mechanism to represent the agent livelihood

1) Basic standards of "livelihood" (element characteristics).

We use the loss probability of livelihood's sufficient and necessary conditions $P(SN)$ to represent livelihood degree, i.e., $f[P(SN)]$ = livelihood degree. Now the expression of f: When $P(SN) = 1, f[P(SN)] = 0$; when $P(SN) = 0$ (to express absolutity), $f[P(SN)] = \infty$ (to express the extremity unachievable in reality); when $P_1(SN) > P_2(SN)$, $f[P(SN)] = \log P(SN)$. Then it can be proved that:

$$f[P(SN)] = \log P(SN)$$

The probability is uncertain, $P(SN)$ = the probability of the loss of sufficient and necessary conditions.

Fuzzy is uncertain, $P(SN)$ = the subordination of the loss of sufficient and necessary conditions.

Information is incomplete and uncertain, $P(SN) =$ the subordination of rough set of the loss of important conditions.

We have noticed that in complex environments the changes of the restraint condition C result in the changes of the sufficient and necessary condition for the existence of the livelihood of the concrete profile function agent. So the livelihood degree f is dynamic and changing with time, and its threshold/valve value should mark the time coordinate. Further study on the relation set of time change rate of f (differential equation) is an effective way to further study the livelihood of agent. In addition, the smaller the livelihood of the sufficient and necessary condition generating negative functions is, the better.

2) The calculation of the livelihood of the composite functions synthesized by single functions.

The law that composites functions are synthesized by single functions, the calculation of the change of the livelihood in the composition of composite parts synthesized by a single function will be conducted.

Suppose $P_1(SN), P_2(SN) \ldots P_N(SN)$ is the failure probability of the necessary and sufficient condition of single functions, then the synthesized livelihood.

$F(\text{synthesis}) = -\log[P1(SN) \times P2(SN) \ldots \times Pn(SN)/\text{absolute necessary and sufficient conditions}]$ ($\times$ indicating some operation law in $P_1(SN) \ldots P_n(SN)$), it is the guideline in the process of the composition of composite functions synthesized by single functions. Suppose a composite function is composed of three functions together, and their respective failure probability of necessary and sufficient condition is $P_1(SN)$, $P_2(SN)$, $P_3(SN)$, their respective failure probability of necessary and sufficient condition for the successful composition of composite functions is $[1-P_1(SN)] \cdot [1-P_2(SN)] \cdot [1-P_3(SN)]$, ($P_1(SN), P_2(SN) \ldots$ exist independently), and the failure probability of necessary and sufficient condition for composite functions is:

$$1 - [1-P_1(SN)][1-P_2(SN)][1-P_3(SN)]$$
$$= P_1(SN) + P_2(SN) + P_3(SN) - P_1(SN)P_2(SN) - P_2(SN)P_3(SN) -$$
$$P_3(SN)P_1(SN) + P_3(SN)P_2(SN)P_1(SN)$$

So the livelihood of the function composed of the three sub-functions is,

$$\log[P_1(SN) + P_2(SN) + P_3(SN) - P_1(SN)P_2(SN) -$$
$$P_2(SN)P_3(SN) - P_3(SN)P_1(SN) + P_3(SN)P_2(SN)P_1(SN)]$$

which is a much smaller livelihood than the single one.

3) The livelihood in the formation of the necessary and sufficient conditions composed of multiple necessary conditions (synthesize the probability of necessary and

Chapter 5 Basics of System Knowledge, Representation and Description

sufficient conditions first).

The completion of each of the necessary conditions, in the sense of physics, is to remove an obstacle in the existence of things, e.g. things are existing in the forms of "presence" and "absence", which will surely survive if all the obstacles are removed. Therefore, the existing yet, complete necessary conditions constitute necessary and sufficient conditions. If the completion of the necessary conditions is represented by the probability, the probability of the necessary and sufficient conditions is the product of the probabilities of all the necessary conditions, that is, $P_1(N) \cdot P_2(N) \ldots P_n(N)$, $P_n(N)$ is a single probability of necessary conditions, and the product of the probability of necessary conditions. As to $F = -\log[1 - P_1(N) \cdot P_2(N) \ldots P_1(N)]$, the more the number of necessary conditions is, the smaller the degree of livelihood becomes.

4) The self-organization mechanism of a single function is formed based on a variety of conditions, and the calculation of the livelihood of the organization mechanism of a single function further forms the complex function.

The calculation procedure can be divided into two parts, i.e. first necessary and sufficient conditions for the self-organization mechanism of a single function based on a variety of supporting conditions are formed, and the sufficient and necessary conditions for the SOM of complicated functions are formed according to the principle that the complex functions are synthesized by single ones (plus the constraints by probability laws), and the complex livelihood can be calculated.

Suppose the necessary and sufficient condition of a single self-organization function is $P(premise\ 1)$, $P(pre\ 2) \ldots P(pre\ n)$, and $P(sup\ 1) \ldots P(sup\ n)$, and supporting conditions are parallel supporting, then $P(SN) = P(pre\ 1) * P(sup\ 1) +$ and $P(pre\ 2) * P(sup\ 2) + \ldots$, but the constraint condition that probability is independent and irrelevant and that the sum of all probability is 1 should be satisfied. The solution for the sufficient and necessary conditions of composite functions based on the principle that the composite functions are synthesized by a single function can be generated, and eventually, the livelihood of the composite functions can be solved.

5) The calculation of livelihood of agents with function living coordination.

The livelihood of agents with function living coordination can be regarded as the composition of two incompatible parts: one is normal working event, and the other is the superposition of events which have been coordinated from the state of function failure to the state of working function. If the livelihood of the two events is represented by probability measures, then it can be expressed as the addition of the probabilities of single events,

$$P \text{ total livelihood} = P \text{ act} + (1-P \text{ act})P \text{ adjustment}$$

The probability of the total livelihood e is P total livelihood after the agents being coordinated via self-livelihood. P livelihood is the original livelihood while P coordination is the success probability of livelihood coordination. See Table 5-1.

Table 5-1 Activity adjusted activity degree

P livelihood	(1−P livelihood)	P coordination	P total livelihood
0.8	0.2	0.7	0.94
0.7	0.3	0.7	0.91
0.6	0.4	0.7	0.88
0.5	0.5	0.7	0.75
0.4	0.6	0.5	0.7
0.3	0.7	0.5	0.7
0.3	0.7	0.8	0.86

From

$$\Delta P_{livetotal} = \frac{\partial P_{livetotal}}{\partial P_{live}}\Delta P_{live} + \frac{\partial P_{livetotal}}{\partial P_{adjustment}}\Delta P_{adjustment} = (1-\Delta P_{adjustment})\Delta P_{live} + (1-\Delta P_{live})\Delta P_{adjustment}$$

we can get that when the livelihood coordination is increased, especially when the original livelihood is quite low, which enhances the efficient functional coordination, the total livelihood of the agent can be greatly promoted.

The living coordination function of agents is made up of three links: acquiring operating information, judging and decision-making, coordination of directive formation and transmission. The combination of hardware and software constitutes the embedded inner part of agents to enhance the livelihood of the agent. This is an important component constituting living agents, and will be further discussed later.

(B) The direct quantitative analyses of the representation of livelihood using living agent functions

The core concept of this approach is that livelihood is expressed by agent functions, and that inferior functions represent low livelihood. As the quantitative representation of living agent function is also classified as "livelihood", another route of a direct expression of livelihood by agent function is created. The establishment of precondition for the working function is usually taken into account in the functional expression of the agent, and the working probability of the agent function can be used to measure that of expressing function, which is a direct way to express the livelihood of the agent function. (Representing the function livelihood with the loss probability of the sufficient and necessary conditions for function livelihood is method to explore the "root and origin", yet the solver is rather complicated. This is a disadvantage.) The examples of this method (due to limited space) will be discussed in Chapter Seven.

5.4.7 Adjustment and Maintenance of the Livelihood of the Function of the Agent and the Multi-Agent System

When the indices of the functional livelihood of agents or multi-agent systems are lower than the required value, measures should be taken to adjust and maintain the functional livelihood.

(1) Basic concepts and principals.

"The adjustment and maintenance of the functional livelihood of the agents and multi-agent systems" is the process to understand, and regulate and control the artificial systems (or future artificial systems) using multi-living agent method to maintain the functions. See Figure 5-8.

1. Display of the Process (expand along the time axis)

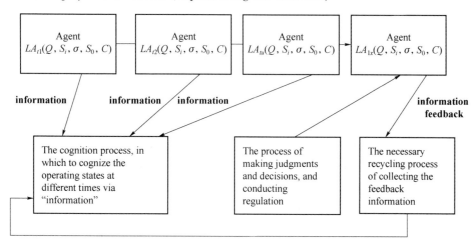

Figure 5-8 The process of unfolding along the time axis

(2) Basics of forming the "maintenance of the livelihood of function".

A. The application of the philosophy of the being of things

An important form of the being of things is the dialectical existence of "universality" and "particularity". The difference between one thing and the other lies in "particularity". The more specific the space and time scales are, the stronger the thing's particularities are. For example, when the parameter of time is second, and the spaces is meter, a person's "birth" has a unique particularity with the two conditions.

In the process of adjusting and maintaining the livelihood of "agent'" functions, we can judge, decide and regulate and control according to the agent state represented by "information" (all kinds of particularities). The particularities of things can be expressed by the "relations" of things of the same kind. The commonly used way is to map the existence space of things into a functional space using various functional spaces and their properties to express the differences of relations of things, e.g., the generalized sizes of things can be described in a normed space, the degree of disparity of things in a distance space, and the direction (angle) relations of things using inner product space. Based on that, we can have a quantitative description on the state and the degree of deviation from standard values, which can be used as a basis for regulation and control.

B. Forming a model combination of the self-organization mechanism of the "agent's" specific function (corresponding to the sub-agent function)

The content is as follows: expand the dynamics expression $LA_i(Q, S_i, \sigma, \delta, S_0, C)$

To make up a group of models used to represent the different inputs and the mechanisms the agent function should have in different environments. At the same time, some models of the model group can be used to correspond with a sub-agent encompassed by the corresponding agent.

This "work" can be not only used in the adjustment and maintenance of the livelihood of "agent" functions but also used in constructing agent functions in the design phase. The structure of the model group is as shown in Figure 5-9.

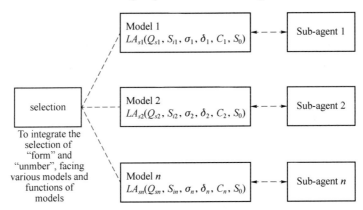

Figure 5-9 Self-organizing mechanism model components

Where $LS_{S1} \ldots LS_{sn}$ represents the dynamics expression of each model in the model group which has completed the intended functions (expressed by S_0). The expressions should have different functional mechanisms to meet the theory that "creating a new order" requires a deviation from equilibrium so that the agent can accomplish the required function under different environments and input conditions. "Selection" is a working process, and is completed in various ways. E.g. the agent design phase consists

Chapter 5 Basics of System Knowledge, Representation and Description

of the designers selecting the model group, the dynamics expressions ($LA_{S1} \ldots LA_{Sn}$) of all models and their inner fine adjustment mechanisms. In the operation of agent, the process of adjusting and maintaining the functions livelihood is completed via the operation of acquiring information and making judgments and decisions according to the setting logic principle and expert system knowledge to form adjustment instructions to adjust functions information. The selection is carried out in two layers, i.e., "form" layer and "parameter" layer. The "form" layer indicates the completion of models, including the mechanism and environmental relations of models, while the "number" layer refers to the selection of the minute parameters of the models, operated usually from top to bottom levels, yet in practice, the intertwined feedbacks are usually needed to finish the "selection".

C. Key elements of "Information"

"Information" is the representation and description of things' motion states. Therefore, based on the knowledge of both the self-organization mechanism of agent functions and the environmental conditions of "agents", relevant information can be acquired and perceived. The information acquired and perceived can help to judge whether the agent operates normally or not, and to find the cause of the abnormality of the agent as well. Based on the judgment, some appropriate measures should be taken because of the adjustment and maintenance of the livelihood of function. The reaction of the agent implementing the "measure" also needs another round of information acquisition and perception, judgment and verification. Therefore, starting from the stage of cognition, "information" has been an indispensable element in the adjustment and maintenance of livelihood of function.

(3) Focus of the adjustment and maintenance of function livelihood.

There are two focuses in the adjustment and maintenance of function livelihood. One is the "healthy" existence and operation of the agent, which can be realized when the mechanism and environment are "healthy", being precautious beforehand. The other is when something is wrong with the function, the problem can be got rid of to recover its normal function (the treatment stage).

As to the first focus, the main part is to analyze if the state of S_i, δ, σ, C is normal, or changes to some extent. If some changes do happen, the consequences testing and the experiments of adjustment measures dealing with the changes should be done. All the accumulated experience can be regarded as "knowledge" to be put in expert system, and can be used in right circumstance.

As to the second focus, we can place extra emphasis on the testing of S_i and S_0, using S_i and its agent functions to test the deviation from normal situation S_0, and conducting fast detection to agent function approaching zero and the feathering function. Some urgent measures to adjust and maintain the functional livelihood should be taken. And in practice, the two types of "regulation and control" could be used in a mixed way with their own characteristics.

(4) Thought and method to judge and make decisions.

Apply the principle of "being mutual and complementary" to a variety of measures to make joint achievements.

Apply the method of acting in a diametrically opposite way, then the two opposing sides complement each other.

Apply the method of acting following the same way, then the two complementing sides oppose each other. See Figure 5-10. Used in a winning occasion in the conflict between the opposite sides, the first part is the action method, and the latter part is the result of the action. the intervention of the spatial-temporal factors cause many consequences—the combination of results(As is shown in Figure 5-10), which will be discussed later.

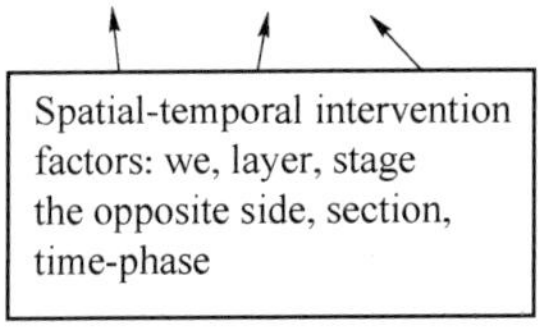

Make concrete and specific judgments and decision using the three imbeddings of "spatial-temporal factors".

(5) The principle of combing hardware and software which constitute the practical platform.

Figure 5-10 Formation of specific judgment decisions using three embeddings of "spatio-temporal factors"

The constitution of the practical platform depends on the scientific combination of hardware and software, in which the hardware has such advantages as solid, safe and reliable, high-speed while the software has great flexibility, and strong universality. Making full use of the two to form an advanced platform is another element of realizing the agent's functional livelihood adjustment.

2. The composition of the agent with the livelihood adjustment mechanism

Based on the concepts and principles described in the preceding section, and combining the dynamic expression (LA_i) the "agent" structure which can help maintain the functional livelihood is formed. Compared with the traditional structure, this structure is made up of three parts, i. e., information perception, judgment and decision-making, and implementation of adjustment instructions. Nowadays, the digital combination of hardware and software is applied quite often. In hardware, the module is embedded into ontology, while software usually refers to some components which can be flexibly injected into the ontology. In the perception of the individual agent, the way of decision-making and adjustment should generally be based on the mode of logical laws + simple expert system (including a simple knowledge base). As "artificial intelligence" develops, the function can be improved gradually. But as this work is mainly an inference of the cause, purpose, and result of the process of things (past and future) under strict restraints to complex unknown things' states, we can conclude according to the philosophy as follows: improvement can be made with the development of science and technology but in a relative way. Certainly, more effective prior information is important in adjusting and maintaining the forming of the operating strategies of agent functions. The newly-constituted agent structure which has the function to adjust and maintain the functional livelihood is shown in Figure 5-11.

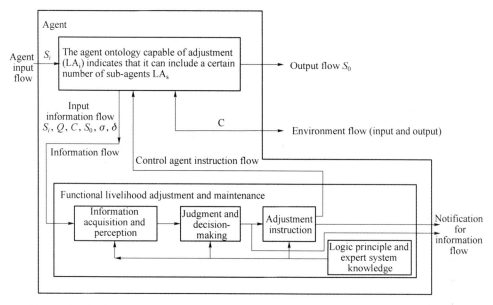

Figure 5-11 Composition of active agent structure with functional activity adjustment and maintenance function

(It is also suitable for agents, and multi-function sub-agent conversion is required.)

(6) Key factors in the realization of the adjustment and maintenance of functional livelihood

Based on what we have discussed above, we will talk about the "key factors in realization" in the following parts. "Realization" is a complicated question in complex conditions. It is more complicated to judge and make decisions when the other side intervenes and creates a countermeasure environment on purpose (Because the opposite side is also adept in the method: acting in a diametrically opposite way, then the two opposing sides complement each other; acting following the same way, then the two complementing sides oppose each other, and can form complex attacking mechanisms.)

A. Link of Information acquisition and perception

The logical relations of this part is shown in Figure 5-12.

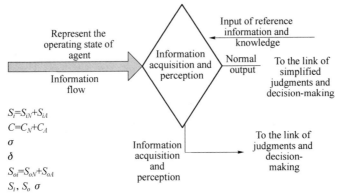

$S_i = S_{iN} + S_{iA}$
$C = C_N + C_A$
σ
δ
$S_{oi} = S_{oN} + S_{oA}$
S_i, S_o, σ

Perception basis: agent operating state information expression
Processing method: according to the specific som and setting parameters of agent function
Input flow property and parameters
Environment flow and its parameters
To compare and extract the disparities between the practical operating state information and setting value (or the Due Value), and process quantitatively the disparity's information (signal) and the correlation degree of information (based on the requirements of livelihood), and then to transfer to the judgment link according to the classification of the results of information acquisition and perception.

Figure 5-12 The working principle of information acquisition perception link

B. The link of decision-making

This link is both important and complicated, in which the importance lies in that the adjustment and maintenance of the livelihood of agent functions are realized through judgment and decision-making, which is rather obvious. But in complicated situations, only the correct decisions make sense, so we must be fully aware of the complexity. The key factors of complexity can be summarized as the following two aspects: Firstly, there are a variety of mechanisms and properties resulting in the loss and change of the livelihood of the agent functions. Generally speaking, the "change and loss" caused by the natural environment and social evolvement are periodical, most of which only require one open ring to adjust the livelihood; whereas the loss and change resulting from human countermeasure purposes and actions mostly have strong fighting strategies. Its lasing property requires that the judgment and decision-making, and adjustment must have excellent performances in space-time domains. Secondly, decision-making is under the influence of various multi-layer spatial factors (satisfying the time restraints simultaneously). In this section, we have discussed the judgment and decision-making in adjusting and maintaining the livelihood of the agent functions under the complex conditions that humans actively plot countermeasure attacks. In this case, the "information" factor has turned into the issue of "countermeasure information" (countermeasure information: the prior information

Chapter 5 Basics of System Knowledge, Representation and Description   117

necessary for countermeasure and that caused by countermeasure actions). In the light that the opposite side employs "reinforced attack measurements", i.e., the combined attack measures with the inducement attack as the guide to induce our side to expose our countermeasure bottom-line, making judgments and decisions to adjust and maintain the livelihood is complex, difficult and the result is uncertain. But efforts should be made to strive for effective "judgment and decision-making". Its block diagram (reducible) is as shown in Figure 5-13.

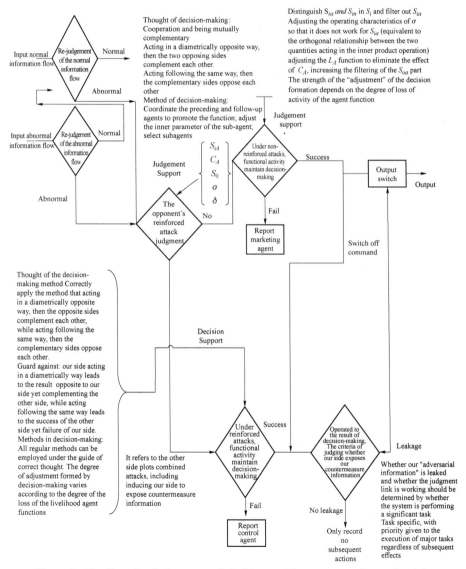

Figure 5-13 Complex judgment and decision-making processes for maintaining and adjusting the living of agents' functions

Many factors are involved in the above-complicated process of making judgments and decisions. In concrete situations, we must make careful research to make the right decision, such as the "focus" on the adjustment of the livelihood of the function;

Whether the other side has adopted reinforced countermeasures and how to discern them;

Thought on the selection of adjustment methods;

The degree of adjustment

The adoption of multiple decision-making methods

The adoption of multiple adjustment methods

....

The results from the judgment and decision-making after considering and selecting the right influencing factors are still uncertain (They may be success or failure.) But regardless of the success or failure of results, we should record the related states and parameters in the process, and then generalize and induce them into "knowledge"for the use of following-up judgment and decision-making. The quality of judgment and decision-making can thus be continuously improved.

5.4.8 The Multi-Layer Realization of the Adjustment and Maintenance of the Livelihood of the System Functions of the Multi-Living Agent

The "realization" is composed of four layers of "adjustment and maintenance of the livelihood of the functions":

The adjustment and maintenance of the livelihood of the functions of the agents themselves—the basic layer

The negation and coordination among agents—the preliminary system layer or enhanced basic layer

The negation and coordination among the agents of the management and control agent group the basic layer of the system

The multi-agent negation and coordination involving both the administrators and the management and control agent group—the enhanced layer of the system

5.4.9 Negotiation and Coordination of the Adjustment and Maintenance of the Livelihood of the Functions Among Multiple Agents

A. Basic norms

1. All "agents" are equal originally at work, with no relations between the highers

Chapter 5 Basics of System Knowledge, Representation and Description

and the lowers.

2. The negotiation and coordination between agents abide by the setting rules (Rules enjoy priorities.)

3. The rule of negotiation-coordination between agents mainly refers to transferring the combination of logic rules and the not-too-complicated expert system "knowledge" according to the setting conditions.

4. In the management layer, all agents obey the instruction of negotiation and coordination between agents given by the management and control agent from the perspective of giving play to the overall functions of the system.

5. All agents report to the management and control agent the operation state and the implementation of "instructions".

6. In the case that a certain agent meets difficulties or contradictions in operating the negotiation-coordination requirements of other agents, it should send out to them the feedback information of failing to complete the negotiation.

B. Realization of negotiation and coordination between agents

1) The basis of the negotiation-coordination mechanism between agents is that all agents have functional living self-organization and livelihood adjustment and maintenance mechanisms.

2) In the case that a certain agent fails to regulate and control the adjustment and maintenance of the livelihood of its functions, and that the problems may be caused by the preceding output, it can deliver cue information of abnormal output to the preceding agent and ask to improve the output.

3) In the case that a certain agent operates abnormally and fails in self-adjustment, it should deliver early warnings to the follow-up agents, and report to the management and control agent simultaneously.

4) In the case that some agents are in the shunt operating mode, and that the operation of one of the agents ceases to be effective, it can ask for assistance from the parallel work agent; the effect of coordination is obvious in operation when the processing workload is too heavy, the individual agents' operation is saturated and blocked, and the functions between agents complement each other.

5) The negotiation-coordination information between agents constitutes "interconnection and interflow" via c input of the environment end.

The following is the brief flowchart of the negotiation and coordination between agents. See Figure 5-14.

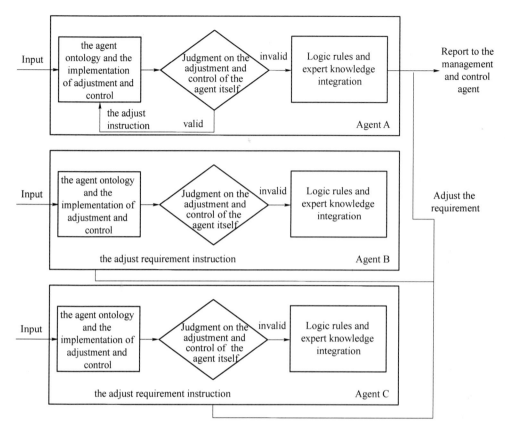

Figure 5-14　Brief process of inter-agent consultation and coordination

5.4.10　Management and Control Agent

1. Tasks undertaken by the management and control agent in a large-scale system.

(1) Regular management tasks in the normal operation of the system

(2) Coordinating the sub-system tasks in normal operation

Normal operation management

Service management (e.g., charging, complaint management, and soliciting suggestions)

(3) Operation maintenance management

(4) Short-term service development (e.g., add service items, or delete out-of-date items)

(5) Long-term service development management (e.g., update the system functions, and expand the lapping functions with other systems)

(6) Other managements

(7) Management of the adjustment and maintenance of the livelihood of the

system functions, which manages and controls the agent kit functions to cooperate with the multi-living agent to constitute new system functions.

2. the work mode of management and control agent.

(1) The centralized management and control mode (largely used in small-scale systems)

(2) The distribution mode (or distribution mode + hierarchical management and control)

(3) Mixed mode: employed in the system's high-level centralized management

The hierarchical and part-type management uses a distribution mode

3. This section discusses the tasks undertaken by the management and control living agent in the adjustment and maintenance of the livelihood of the system functions in the multi-living agent system.

Multi-living agents adopt the mixed mode of management and control to adjust and maintain the livelihood of the function, which has been specified from four layers in 5.3.8.

Management and control agent in the base of distribution hierarchical management and control is mainly responsible for the basic adjustment and control of system and assists the administrator to manage and control the adjustment and maintenance of the livelihood of the function in the system enhanced layer (It's part of the management and control work of the whole system.)

4. The specific management and control work of management and control.

In the case that agents cannot reach an agreement on the necessary negotiation and coordination, the management and control agent does negotiation and management and control at the system level.

> In the case that the management and control agent's management and control and negotiation fail, report to the administrator.
> Assist the administrator to manage and control the system
> Rectify the negotiation results of the agent or between agents in the system function layer
> Receive from all agents the report of the key operations, and carry out key monitoring of system operation

5. General description of the mechanism of the adjustment of the livelihood of function of the management and control agent.

The previous sections discussed the agent's composition and multi-agent system structure using the agent dynamic expression $LA(\ldots)$, two-set model, and three-set model, highlighting the concept of livelihood, while my other academic papers expand the discussion of information systems and the working principle of constituting agents. In this section, the generalized work mechanism of the management and control agent will be discussed. The constitution of the mechanism is an application of the relations mapping inversion method. See Figure 5-15.

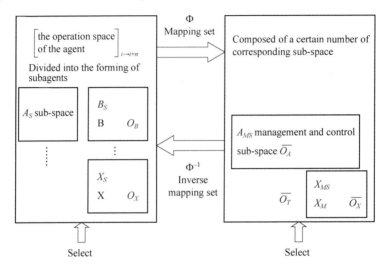

Figure 5-15 Generalized diagram of the functional activity adjustment mechanism

Where $A, B, \ldots X$ represents the relevant agents and their constitutions (It can also represent sub-agents.), the operations of O_A, O_B, O_X represent the interfunction within the agents (sub-agents). So A_S, B_S, X_S space represents the operation of agents (sub-agents), while the time interval $i \rightarrow i + n$ represents the management and control agent in the operation process of a certain time period. The management and control of the adjustment and maintenance of the livelihood of the functions of the agent is operated in the system layer, the operation of the space in which the agent operates maps to the management and control space of the management control agent via corresponding information, while in the management and control space coordination and control are formed via "selection" (i.e., the process of implementing information perception and making judgment and decision), new structures of agents (sub-agents) are formed, which can be expressed with the space formula $A_{MS} \ldots X_{MS}$, where $\overline{O_M} \ldots \overline{O_M}$ represents new operation rules; the anti-mapping Φ^{-1} indicates to transfer $A_{MS} \ldots X_{MS}$ to the agent's operation space and make an adjustment to form a new livelihood of the function of the agent; while the selection of the agent's operation space represents the initial design or setting.

6. Structure of the management and control living agent and Realization of the coordination and control of the livelihood.

This section is an expanded discussion of the preceding section.

1. Structural composition.

The general structure described in Section 5.3.7.B.1 can still be used here, only that the characteristics of the composition of the agent ontology must be explained as shown in Figure 5-16.

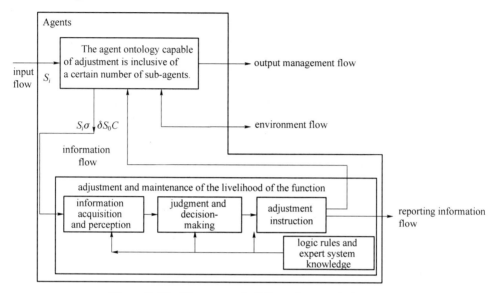

Figure 5-16 Management of the active agent structure

The agent ontology is mainly composed of the algorithm that constitutes the management and control function, the software modules, and the hardware platform which implements the operation. The algorithm and software modules call the instruction in control to adjust and control, and form the adjustment and control instruction to deliver to all agents (different from ordinary agent functions). The input flow mainly refers to various information flows, while the decision-making (logic rules and expert system knowledge) in the adjustment and control of the livelihood attaches much importance to making rules via the consideration that whether the whole system functions well, the content of the rules and the expert system knowledge are more complicated than those of the individual agent, and they also increase much in quantity. In future development more advanced artificial intellectual technologies will be applied. This section thus ends with the emphasis that ensuring the livelihood of the function of the management and control agent is the basic prerequisite, and that consequently "judgment and decision-making" should employ various methods to judge and decide whether the management and control function of the management and control agent work normally or not, e.g., embed the known input flow, and then comprehensively judge whether it functions normally or not via "output flow".

2. Realization of the management and control of the adjustment and maintenance of the livelihood of the system function via the management and control agent.

For the constitution and the block diagram. The characteristics are as follows:

The content of the judgment and decision-making block diagram can be tailored according to the practical situations. Timely and conscientious judgment and decision about the livelihood of the management and control agent's function should be made. "Judgment and decision-making" lay more stress on the functioning of the system, and the judgment that entails setbacks to the whole system in which "The local succeeds while the general fails", "The short-term succeeds while the long-term fails" should be energetically avoided.

E. g. under the opposite side's inducement or exploratory attacks, high-level countermeasures are resorted to, or the entire measurements are exposed to the opposite side; when discovering a new attack measure, too quick use of the related measures which can directly eliminate the attack consequence yet ignoring mastering the attack's important mechanism will result in difficulties in in-depth effective countermeasures (similar to trying to treat the patient with a fever by using a quick and heavy dose of antipyretic). Rectifying the fact that the agent layer takes countermeasures obstructs the more effective results in the system layer.

Management agents, in the adjustment and maintenance of the livelihood of the system function, make"ineffectiveness" judgment based on facts, which is an "opposite" result directly perceived through the senses, yet it promotes the administrator to make decision of realizing the new structure of the agent system. This is the starting point of a new "opposing each other yet deciding each other" effect.

5. 4. 11 A Theoretical Comparison Between the Dissipative Self-Organization Theory (Key Content) and the MLAM Theory

Dissipative self-organization theory is an important component of modern system theories. It extends the characteristics of the research on physics from the state of motion to the process of motion, noticing that the motion of the contradictions is the key factor in the process of moving. The dissipative self-organization theory is an important theory in the basic layer of system sciences, with the obvious characteristics of universal basic theory, while the multi-living agent method is a theory and method representing and designing the system based on the dissipative self-organization theory, which actually extends into the application basic layer and application layer. The MLAM shares the same importance with the concept of "application", the extended comparison is as shown in Table 5-2.

Table 5-2 Theoretical comparison of dissipative self-organization theory and multi-active agent approach

	Dissipative self-organization theory	Multi-living agent method and theory
Description of the entire system dynamic self-organization characteristics	The prerequisite for the system's order survival is the existence of the open dissipative structure.	Extend to: with the livelihood of the individual agent as the basis, multiple living agents constitute multi-living agents, which represent the self-organization order motion of the system.
	The dissipative has the entropy necessarily produced by irreversible motion, which stops the continuous multiplication of the entropy included in the system. This is the basic condition for the order motion.	The maintenance of the self-organization mechanism of the livelihood of the functions of all agents embodies the order motion of the agents.
	Only that the system is far from the sate of balance, new "orders" can occur.	In a multi-living agent system, the individual agent runs the self-organization mechanism of the livelihood of the functions, making maintenance and adjustment, and doing negotiation and coordination between agents (including the adjustment and control done by the management and control agent). Most of the work is the essential adjustment to the original state.
The evolution of the system and the co-evolution of the small and large universes.	The system achieves new order via landing.	In the self-organization adjustment and maintenance of the livelihood of the function, it's displayed in achieving new order in the process of "landing".
	Co-evolution of the small and large universes	When the environment (large universe) changes, the adjustment and maintenance of the livelihood of the functions in the multi-living agent system reflect co-evolution.
	Evolution of the evolution mechanism	The proposal and application of the multi-living agent method and theory improves the method of representing "complex" artificial systems, and up-grades the design method and service function of artificial systems. It's a development of the development mechanism.

Continued

	Dissipative self-organization theory	Multi-living agent method and theory
The system self-organization mechanism is destroyed, and the severely destroyed will secede permanently from society.	No dedicated description is only included in relevant principles. The continuously produced entropy must be dissipated, and both the small and large universes co-evolve.	Expressed using the livelihood of the agent, especially the adjustment and maintenance of the self-organization livelihood of the function. It's the focus of the research and practices to realize the extension of artificial systems' service term as long as possible.
The framework by Professor Qian Xuesen	Used to explore the complex giant system theory and the methodology: e. g., the combination of the qualitative and quantitative methods, the "integrated discussion hall" method which gives play to the experts, and the human-machine method	Research and practices on the method and theory, e.g., in multi-living agents constituting the system structure, the management, and control agent are established to store the related wisdom gathered from the design stage and the designer community to enable the system to continuously function in a complex environment, and the service term can thus be extended.

5.4.12 Exemplifications of Applications

Case 1: Construct the maintenance of the livelihood of the function of the local communication system

In designing a communication information system in local areas, aside from considering satisfying the requirement of users during peacetimes at the top level, we should also consider the additional special services of this system during emergency times (e.g., power off). Consequently, in general, the system can be simplified into two layers: agents at the system level and conditional sets of agents. The system's backbone layer consists of the fiber-optical telecom network satellite communication agent, the satellite communication agent, and the mobile communication network agent, while the emergency backup agent consists of a low-power shortwave transmitter-receiver. See Figure 5-17.

The small shortwave transmitter-receiver is cheap, easy to operate, contains the antenna, has high sensitivity, can operate at long distances due to reflection by the ionosphere (F layer), and can also operate from power generated by a hand-operated electrical generator. Hence, it can still operate in the absence of electrical power, shortage of oil, and when the optical fiber is down. For example, in earthquake areas, when the electrical network has an emergency failure, then the livelihood of the optical fiber

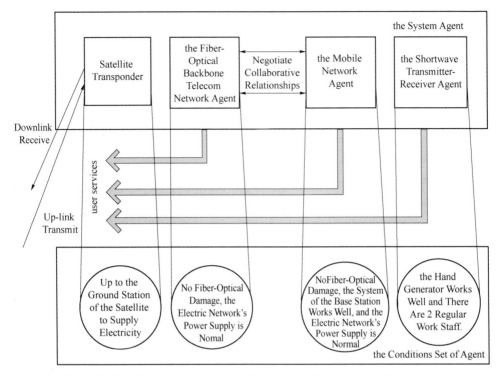

Figure 5-17 The system structure of communication system in local areas.

network and the network of the mobile communication system are lost and cannot fulfill the task. At this time, the shortwave transmitter-receiver can be used to meet the emergency requirement and fulfill the work. This fact shows the substitution relationship among the agents, and also tells us that it is not necessary for all of the agents to be advanced when designing the information system in the complex situations. Consideration of the existence of the system livelihood should be based on the specific situations of the system. E. g., Premier Wen Jiaobao immediately flew to the first front of Sichuan earthquake area 2 or 3 hours after the occurrence of the quake to conduct the work of providing disaster relief. But due to the fact that the power supply and the optical fiber in the disaster area broke off, the communication in the quake epicenter broke off, too and Premier Wen's conduct work was thus affected. If the quake epicenter had constructed emergency agents of shortwave transmitter-receivers, earlier and more information about the epicenter could have been obtained even under the condition that the electricity and mobile communication both broke off.

From the case discussed above, we conclude that setting up the concept of livelihood of the system functions has scientific and practicality. Furthermore, the application of the multi-living agent method greatly improves the quality and efficiency of artificial system design. This can be regarded as an evolution of the artificial system design mechanism.

Case 2: Research on network distributed intrusion detection system using the multi-living agent

(1) Introduction

The openness of the Internet facilitates information sharing and exchanging which seriously challenges the information security. Information security has evolved into a critical issue of information system. Intrusion detection technology, as an active information security safeguards, can effectively remove the defects of the traditional security protection technology while the traditional intrusion detection system can hardly satisfy the security requirement due to the wide application of distributed computing environment and large-scale network. Thus, it is necessary to share the information and detect the intrusion collaboratively between each intrusion detection system which promotes the birth and development of the distributed intrusion detection system. Consequently, the distributed intrusion detection system has eventually turned into research focus of intrusion detection and even the entire field of network security.

The earliest distributed intrusion detection system was proposed by snap et al. in 1991[2], and 'Distributed collection, distributed processing, centralized management' is the core of distributed intrusion detection. This distributed detection method is effective to detect the large-scale attacks and distributed attacks. However, this method would definitely make the console overloaded, and the network would also become overloaded when the nodes send mass data to the console, which would ultimately affect the efficiency and accuracy of the intrusion detection system [3, 4]. For solving these problems in distributed intrusion detection system, Kruegel et. al. proposed the distributed intrusion detection system using mobile agents in 2001. In this way, the console only manages the mobile agents while all of the network data is detected by the mobile agents. The network load and computation of the console would be dramatically relieved. Nevertheless, this will lead serious security risk in the distributed intrusion detection system using mobile agents, i.e., the entire system would lose the detection capability when the mobile agents are attacked and fail to work. Therefore, the security of mobile agents has turned into a hot issue in the relevant research.

To improve the security of distributed intrusion detection systems based on mobile agents, this paper first considers the quantitative analysis method of agent livelihood in the distributed intrusion detection system, which is used to obtain the changes in agent livelihood state value and the key parameters impacting the livelihood status. Then we propose a way to realize the living agent based on the mobile agent technology, which is according to the changes of agent livelihood state value in a different moment. Finally, we propose a novel framework for a distributed intrusion detection system, namely the distributed intrusion detection system based on the multi-living agent (DIDS-MLA). In this novel system, the living agent is the basic unit of the system, and DIDS-MLA expands the traditional DIDS from a two-layer structure to a three-layer structure by adding the livelihood state monitoring agent. In this way, DIDS-MLA can monitor the living state value of the detection agent deployed at the key location in the network and can adjust time according to the changes in agent livelihood state value. Thereby, DIDS-MLA can ensure its stability of itself when the detection agents lose their livelihood being attacked. Finally, we can improve the security and stability of the system.

(2) Research on quantitative analysis method for agent livelihood in multi-living agent

In this section, we propose a quantitative analysis method for agent livelihood in the distributed intrusion detection system according to the dynamic representation of living agents and the knowledge of the queuing theory.

We call the perception of δ and S_A (the attack flow) in $S_i = S_i(S_N \times S_A)$, the perception of δ and C_A (the reverse constraints of attack formation) in $C_t = C_t(C_N \times C_A)$ and the perception of σ forming $\sigma[S_i(S_N \times S_A), C(C_N + C_A), Q_t] = S_{oN}$ as the methods of the agent keeping livelihood for eliminating the intrusion attack. In this process, the agent is considered as the stochastic service system. This system will confront the intrusion service formed by the intrusion flow. If the arrived intrusion flow can not be accepted, the intrusion will be penetrated (successful intrusion). Therefore we can utilize the penetration probability of the stochastic service system to represent the livelihood of the agent reversely. For deep analysis of livelihood of the agent, we divide the intrusion detection agent into two service blocks which is shown in Figure 5-18.

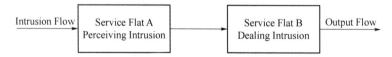

Figure 5-18　Stochastic service system model of intrusion detection agent

We suppose that the intrusion flow is the Possion flow with the parameter λ. The probability of k intrusion flows arrived in the period of τ can be expressed as $P_k(\tau) = \dfrac{(\lambda \tau)^k}{k!} e^{-\lambda \tau}$ where λ is the flow parameter and its unit is the number per unit time. So we can obtain the probability of one intrusion arrived in period Δt:

$$P_1(\Delta t) = \lambda \Delta t(1-\lambda \Delta t) = \lambda \Delta t - \lambda^2 \Delta t^2, \text{when } \Delta t \to 0, P_1(\Delta t) = \lambda \Delta t$$

Furthermore, we suppose that the service time of each service block is a random number, and the probability of the service time obeys the negative exponential distribution. Then, the probability of service time is less than t is:

$$P(t_{service} < t) = 1 - e^{-\mu t} \tag{1}$$

So, if Δt is very small, the probability of service completed in the period of Δt is:

$$P_1(t_{service} < \Delta t) = 1 - e^{-\mu \Delta t} = 1 - (1-\mu \Delta t) = \mu \Delta t \tag{2}$$

So, the probability of the service block A and the service block B completed in the period of Δt time is $\mu_1 \Delta t$ and $\mu_2 \Delta t$.

As we have supposed that the structure of the service block is serial, so the intrusion flow goes through service block A, the output flow of service block A goes into service block B (the output flow could approximate the Poisson flow), and there is no interval between service block A and service block B. In the followings, we will research the living agent through the differential equation of each state probability in the service system.

We suppose that A_{00}, A_{10}, A_{01}, A_{11} respectively represent the state of service block A and service block B are idle, block A is busy and block B is idle, block A is idle and block B is busy, and block A and block B are busy. Now we analyze the differential equation of each state and suppose that the intrusion flow arriving at one service block will penetrate when this service block is working: in the state of A_{00}, A_{10}, A_{01}, A_{11}. We suppose there is penetration.

The probability $P_{00}(t + \Delta t)$ of $A_{00}(t + \Delta t)$ is determined by the following incompatible events:

The original state is A_{00}, and there is no flow arriving. For this case, the probability is

$$P_{00}(t+\Delta t) = P_{00}(t)(1-\lambda \Delta t) \tag{3}$$

The original state is A_{01}, and B has finished service in the period of Δt time. In this case, the probability is

$$P_{00}(t+\Delta t) = P_{01}(t)\mu_2 \Delta t \tag{4}$$

Therefore, we can derive that

$$P_{00}(t+\Delta t) = P_{00}(t) - \lambda P_{00}(t)\Delta t + P_{01}(t)\mu_2 \Delta t \tag{5}$$

and we have

$$\frac{dP_{00}(t)}{dt} = -\lambda P_{00}(t) + P_{01}(t)\mu_2 \tag{6}$$

Chapter 5 Basics of System Knowledge, Representation and Description

1) The probability $P_{10}(t+\Delta t)$ of the event $A_{10}(t+\Delta t)$ is determined by the following incompatible things:

a) The original state is A_{10} and in the period of Δt, the service block A has not finished the service and A is keeping the state of A_{10}, no matter whether there is new flow arriving. The probability of this case is

$$P_{10}(t + \Delta t) = P_{10}(t)(1 - \mu_1 \Delta t) \tag{7}$$

b) The original state is A_{00}, and there is new flow arriving in the period of Δt. The probability of this case is

$$P_{10}(t + \Delta t) = P_{00}(t)\lambda \Delta t \tag{8}$$

c) The original state is A_{11} and in the period of Δt if the service block B has finished service and A has not finished service, the state A_{11} would change into A_{10} whether there is new flow arriving or not. The probability of this case is

$$P_{10}(t + \Delta t) = P_{11}(t)(1 - \mu_1 \Delta t)\mu_2 \Delta t \doteq P_{11}(t)\mu_2 \Delta t \tag{9}$$

d) The original state is A_{01} and in the period of Δt if the service block B has finished service and the state is A_{01} yet along with the new flow arriving. The probability of this case is

$$P_{10}(t + \Delta t) = P_{01}(t)\mu_2 \Delta t \lambda \Delta t \doteq 0 \tag{10}$$

Thus, we have

$$P_{10}(t + \Delta t) = P_{10}(t)(1 - \mu_1 \Delta t) + P_{00}(t)\lambda \Delta t + P_{11}(t)\mu_2 \Delta t \tag{11}$$

which can be further written as

$$\frac{dP_{10}(t)}{dt} = -\mu_1 P_{10}(t) + \lambda P_{00}(t) + \mu_2 P_{11}(t) \tag{12}$$

2) The probability $P_{01}(t + \Delta t)$ of $A_{01}(t + \Delta t)$ is determined by the following incompatible things.

a) The original state is A_{01}. We keep the state of A_{01} to the time of $t + \Delta t$, and at this time the service block B has not finished service without new flow arriving. For this case, we have

$$P_{01}(t + \Delta t) = P_{01}(t)(1 - \lambda \Delta t)(1 - \mu_2 \Delta t) \tag{13}$$

b) The original state is A_{10}. In the period of Δt, the service block A has finished service. At the same time, there is no new flow arriving. The probability of this case is:

$$P_{01}(t + \Delta t) = P_{10}(t)(1 - \lambda \Delta t)\mu_1 \Delta t \doteq P_{10}(t)\mu_1 \Delta t \tag{14}$$

c) The original state is A_{11}. In the period of Δt, the service block A has finished service without new flow arriving, and whether B has finished service or not. For this case, the probability is:

$$P_{01}(t + \Delta t) = P_{11}(t)(1 - \lambda \Delta t)\mu_1 \Delta t \doteq P_{11}(t)\mu_1 \Delta t \tag{15}$$

Thus, $P_{01}(t + \Delta t) = P_{01}(t)(1 - \lambda \Delta t)(1 - \mu_2 \Delta t) + P_{10}(t)\mu_1 \Delta t + P_{11}(t)\mu_1 \Delta t$ (16)

$$\frac{dP_{01}(t)}{dt} = \mu_1 P_{10}(t) - (\mu_2 + \lambda) P_{01}(t) + \mu_1 P_{11}(t) \tag{17}$$

3) The probability $P_{11}(t + \Delta t)$ of $A_{11}(t + \Delta t)$ is determined by the following incompatible things.

a) The original state is A_{11}. In the period of Δt, both the service block A and B have not finished service whether there is new flow arriving or not. For this case, the probability is

$$\begin{aligned} P_{11}(t + \Delta t) &= P_{11}(t)(1 - \mu_1 \Delta t)(1 - \mu_2 \Delta t) \\ &= P_{11}(t) - (\mu_1 + \mu_2) P_{11}(t) \Delta t \end{aligned} \tag{18}$$

b) The original state is A_{01}. In the period of Δt, the service block B has not finished service, and there is new flow arriving. The probability is

$$P_{11}(t + \Delta t) = P_{01}(t)\lambda \Delta t(1 - \mu_2 \Delta t) \doteq P_{01}(t)\lambda \Delta t \tag{19}$$

c) The original state is A_{10}. In the period of Δt, the service block has finished service, and there is new flow arriving. The probability of this case is

$$P_{11}(t + \Delta t) = P_{10}(t)\lambda \Delta t(\mu_1 \Delta t) \doteq 0 \tag{20}$$

Thus,

$$P_{11}(t + \Delta t) = P_{11}(t) - (\mu_1 + \mu_2)P_{11}(t) + P_{01}(t)\lambda \Delta t \tag{21}$$

and

$$\frac{dP_{11}(t)}{dt} = -(\mu_1 + \mu_2)P_{11}(t) + P_{01}(t)\lambda \tag{22}$$

We could obtain the differential equations of the service system state according to the above equations

$$B = \begin{cases} P'_{00}(t) = -\lambda P_{00}(t) + \mu_2 P_{01}(t) \\ P'_{10}(t) = \lambda P_{00}(t) - \mu_1 P_{10}(t) + \mu_2 P_{11}(t) \\ P'_{01}(t) = \mu_1 P_{10}(t) - (\lambda + \mu_2)P_{01}(t) + \mu_1 P_{11}(t) \\ P'_{11}(t) = \lambda P_{01}(t) - (\mu_1 + \mu_2)P_{11}(t) \end{cases} \tag{23}$$

Through solving the above differential equations under certain conditions, we obtain the probability of each state $P_{xy}(t)\left(\begin{matrix} x \\ y \end{matrix} = 1, 0\right)$. Through this way, we could obtain the quantitative analysis method of agent livelihood. In this paper, we are interested in the steady-state solution of $P_{xy}(t)$, namely $\lim_{t \to \infty} P_{xy}(t) = P_{xy}$. It also is the solution of homogeneous algebraic equations when $P'_{xy}(t) = 0$.

In this situation, we need to derive the penetration probability P_{break} and we can verify that the coefficient determinant of B equations is 0. And P_{xy} has the non-zero solution. If $\sum P_{xy} = 1$ (the total probability is 1), we have

$$C = \begin{cases} P_{00} = \dfrac{\mu_1 \mu_2}{(\mu_1 + \lambda)(\mu_2 + \lambda)} \\ P_{10} = \dfrac{\lambda \mu_2 (\mu_1 + \mu_2 + \lambda)}{(\mu_1 + \mu_2)(\mu_1 + \lambda)(\mu_2 + \lambda)} \\ P_{01} = \dfrac{\lambda \mu_1}{(\mu_1 + \lambda)(\mu_2 + \lambda)} \\ P_{11} = \dfrac{\mu_1 \lambda^2}{(\mu_1 + \lambda)(\mu_2 + \lambda)(\mu_1 + \mu_2)} \end{cases} \quad (25)$$

Next, we need to utilize equation (25) to solve P_{break}. The penetration probability P_{break} that the service block B could not service the new flow

$$\begin{aligned} P_{break} &= 1 - \dfrac{\mu_2}{\lambda}(P_{01} + P_{11}) \\ &= 1 - \dfrac{\mu_1 \mu_2 (\lambda + \mu_1 + \mu_2)}{(\lambda + \mu_1)(\lambda + \mu_2)(\mu_1 + \mu_2)} \end{aligned} \quad (26)$$

Thus, we have

$$P_{SuccedB} = \dfrac{\mu_2}{\lambda}(P_{01} + P_{11}) \quad (27)$$

According to the symmetry principle, we could obtain the total probability of the service block A

$$P_{SuccedA} = \dfrac{\mu_1}{\lambda}(P_{10} + P_{11}) \quad (28)$$

The total penetration probability of the stochastic service system considering the intrusion flow penetrate the service block A is

$$1 - \dfrac{\mu_1}{\lambda}(P_{10} + P_{11}) \cdot \dfrac{\mu_2}{\lambda}(P_{01} + P_{11}) \quad (29)$$

Therefore we can obtain the corresponding representation of livelihood

$$P_{living} = \dfrac{\mu_1}{\lambda}(P_{10} + P_{11}) \cdot \dfrac{\mu_2}{\lambda}(P_{01} + P_{11}) \quad (30)$$

After obtaining the quantitative description of living agent, we can further explain the quantitative description for living agent according to the real numerical result. The probability with different values of parameters is shown in Table 5-3.

Table 5-3 Actual numerical culculation reswlts

	P_{10}	P_{01}	P_{11}
$\mu_1 = \mu_2 = 10, \lambda = 1$	0.086	0.082 6	0.004 13
$\mu_1 = \mu_2 = 5, \lambda = 1$	0.152	0.138	0.013 8
$\mu_1 = 5, \mu_2 = 1, \lambda = 1$	0.097	0.416	0.069 4
$\mu_1 = \mu_2 = 5, \lambda = 3$	0.304	0.234	0.071
$\mu_1 = \mu_2 = 10, \lambda = 10$	0.375	0.25	0.125
$\mu_1 = \mu_2 = 1, \lambda = 10$	0.49	0.082 6	0.413

In Table 5-4, we can see that the agent livelihood state value decreased gradually from 1 to 0. When agent livelihood state value reduced to 0, the penetration probability of attack stream reached to 100 percent. This means that the detection function of agent is a failure completely. We can also see that when parameter λ increased, the agent livelihood state value would decrease rapidly. In the distributed intrusion detection system, the physical meaning of parameter λ is the traffic of the network attack stream processed by detection agent in a certain period of time. Namely, we can judge that the agent livelihood state value will decrease rapidly when the network stream processed greatly by an agent in the system in a certain period of time.

Table 5-4 Calculation results of penetration probability and living—agent state values

	$P_{SuccedA}$	$P_{SuccedB}$	P_{living}	P_{break}
$\mu_1 = \mu_2 = 10, \lambda = 1$	$10(0.086 + 0.004\ 13) = 0.9$	$10(0.082\ 6 + 0.004\ 13) = 0.86$	0.774	0.226
$\mu_1 = \mu_2 = 5, \lambda = 1$	$5(0.152 + 0.013\ 8) = 0.829$	$5(0.138 + 0.013\ 8) = 0.759$	0.630	0.370
$\mu_1 = 5, \mu_2 = 1, \lambda = 1$	$5(0.097 + 0.069\ 4) = 0.832$	$1(0.416 + 0.069\ 4)0.485$	0.404	0.596
$\mu_1 = \mu_2 = 5, \lambda = 3$	$5(0.304 + 0.071)/3 = 0.625$	$5(0.234 + 0.071)/3 = 0.508$	0.318	0.682
$\mu_1 = \mu_2 = 10, \lambda = 10$	$1(0.375 + 0.125) = 0.5$	$1(0.250.125) = 0.375$	0.175	0.825
$\mu_1 = \mu_2 = 1, \lambda = 10$	$0.1(0.49 + 0.413) = 0.090\ 3$	$0.1(0.082\ 6 + 0.413) = 0.049$	0.005	0.995

For avoiding the failure statement of agent detection, we need to design a novel framework of distributed intrusion detection system to monitor the agent livelihood status in time. In the Table 5-4, we can see that when the value of parameter λ increased from 1 to 3, the agent livelihood state value decreased. The decline of the agent livelihood state value will influence the detection effects of the system. In this situation, we apply the negotiation-coordination mechanism of living agent into the detection agent. That is, when the performance of detection agent in the system decreased, the system can still work normally through the assistance from the neighbor detection agent. If the value of λ raised up to 10 in a certain period, this agent can't work well. Then the system will create a new agent to replace the failure agent to make sure that the system still works normally. In next section, we will first improve the mobile agent according to the changes of livelihood state value, and then propose the realization of living agent based on the negotiation-coordination mechanism of multi-living agent method which is the base of constructing the novel distributed intrusion detection system.See Figure 5-19.

(3) Constitution of multi-living agent

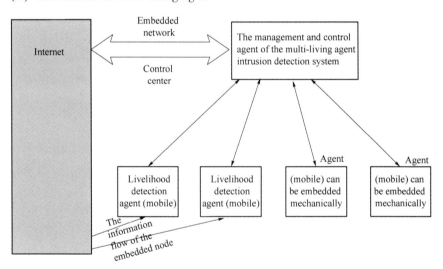

Figure 5-19　Multi-active agent intrusion detection system architecture

(4) Simulation verification

To verify the validity of the distributed intrusion detection system based on multi-living agents proposed in this paper, we utilize the DDoS1.0 in the DARPA2000 of Lincoln laboratory as the test dataset, which includes a complete DDoS attack sequence.

In the DIDS, when the detection agent worked abnormally due to an attack, the control node in the DIDS could not monitor the network. In this situation, the detection system may be unable to detect the DDoS attack due to the unavailability of information on DDoS. Same as DIDS, when the mobile agent deployed in the DIDS-MA could not work usually due to being attacked, the network monitored by the mobile agent would be unprotected. It would reduce the performance of the detection system. Contrarily, in the DIDS-MLA, when the attacking data flow of DDoS arrived at the network being monitored by the detection system, the livelihood state monitoring agent could monitor the livelihood state of the detection agent according to the changes in network data traffic. If the monitoring agent found the detection agent in the state of deactivation or half-deactivation, the system would support it by using the negotiation-coordination mechanism of the multi-living agent. A new detection living agent would be created quickly and replace the failed one. Thus the system could recover rapidly.

To ensure the normality of the detection agent in the system, we could set a threshold of agent livelihood state according to the need of the system detection effect. If the value of the agent's livelihood state is less than the threshold value, the system would create a new living agent to replace the failed one. The comparison between DIDS-MA and DIDS-MLA is shown below.

From Figure 5-20 and Figure 5-21, we can see that the detection agent would unable to detect intrusion when the agent lost the livelihood when DIDS-MA was being attacked. However, the DIDS-MLA, proposed in this paper could create a new agent to replace the failed one under attack and recover rapidly. It can be seen that DIDS-MLA proposed in this paper could protect the agents in the distributed intrusion detection system and could work normally under network attacks.

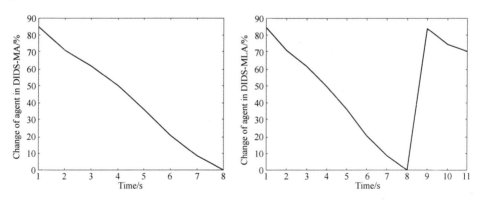

Figure 5-20　Change of agent in DIDS-MA　　Figure 5-21　Change of agent in DIDS-MLA

(5) Conclusion

Case 2 means that the multi-living agent method is an effective way to construct a complex information system. The multi-living agent method could provide support to distributed intrusion detection from the perspective of the top-level. In chapter 7 we will give more examples. This application is elementary, we expect more examples to complete the theory.

Chapter 6
Development and Categorization of Artificial System Design and the Philosophy of Artificial System Design

6.1 Discussions on the Essentials of Artificial System

6.1.1 Discussions on the Connotations of Artificial System

6.1.1.1 Artificial System

Artificial system is a collective term that encompasses all kinds of mainly artificially manufactured aggregates with systematic features and functions that serve human beings, basically to highlight the connotations of the two sections of "artificial" and "system".

6.1.1.2 The Advent and Constant Development of Artificial System

The advent and constant development of artificial system demonstrate the evolution of mankind and the progress of civilization. It acutely exemplifies the merging of "knowledge" and "practice" of human beings, their increasing involvements and understanding of the systematic nature of complex things as well as their accessible achievements affairs in the process of the development of the unity of opposites.

6.1.1.3 Important Artificial Systems

Important artificial systems are characterized by multilayered openness of diverse properties, the success of whose development and design is concerned with social progress, the wisdom and abilities of designers and the organic joint support from the fields of science, technology and engineering of human civilization.

6.1.1.4 Various Artificial Systems Are Involved with Various Fields of Science, Technology and Engineering

Various artificial systems are involved with various fields of science, technology and engineering with three interlinking fields: system science, information science and control science.

6.1.1.5 Artificial System Is Connection of Engineering Science and Technology and Engineering Implementation

As to the learning and practicing abilities of human beings in the development of artificial systems, laws acquired by scientific discoveries are first extended to the fields of technical science (scientific laws that cope with the key developments of technology in application and the support of new core methods) that closely combines science and technology, which are then extended to specialized technological areas of applied fields to form the groundwork for the development of artificial design. The research and development of each specific artificial system are appropriate and effective steps taken to fully play the service functions of the artificial system (The more advanced and diversified, the better.) On the above-mentioned basis with considerations of actual conditions and constraints. This is the main connotation of the fields of engineering scientific technology and engineering construction.

6.1.1.6 The Incessant Joint Evolution of Artificial Systems and Human Society

Owing to the incessant joint evolution of artificial systems and human society, the combination between artificial systems and mankind increasingly tightens. Thus the characters of the "Open Complex Giant System" are becoming enhanced and conspicuous on daily basis, under the umbrage of whose implication the developments of system, information and control sciences cannot meet the demands of the "requirement". On account of the above-illustrated realities, chances of failure have proliferated in the research and development of important new artificial systems, while boosted by the impetus of surging demands, the constant development of artificial systems becomes an inevitable and inexorable trend. These above-elucidated contradictions will persist in the long term, for which conclusive panacea solutions won't be found once for all. They are an embodiment of the philosophic truth of the evolution and development of human society, only to be dealt with through continuous learning and practice in the process of social development.

6.1.2 Discussions on the Concise Categorization of Artificial System from the Perspective of Functions and Characteristics

6.1.2.1 Scientific Type (Scientific Engineering Type)

It is characterized by its purpose for the exploration of scientific laws. In consequence of the fact that those explorations more than often take place in extreme environments, the preconditions for the laws themselves make the artificial systems employed in their exploration very complex with the comprehensiveness of engineering, which are therefore categorized as scientific engineering type. For example, Hubble Space Telescope is a giant complex artificial system of scientific engineering type, the research and development criterion of which is to accomplish planned scientific missions within the limits of budgets. Such as: Antarctic scientific exploration stations, nuclear physics laboratory centers, etc.

Direct economic interests are seldom considered in the scientific engineering type, which depends upon supports in accordance with the significance of those scientific projects (scientific experiments included) in comprehensive national and social development.

6.1.2.2 Public Service Type

In order to fulfill their pubic duties, an assortment of artificial systems is needed by the nation and the government as tools and means to perform public services. These systems will become increasingly colossal and complex.

This category of artificial systems can be further divided into two subcategories: National security type and civil public service type. Such as: military reconnaissance satellites employed in national defense, air defense system, strategic missile system, coastal defense system, military navigation system, and other typical systems applied for national security …

Administrative management system, immigration management system, customs management system, legal service system, taxation management system, national bank management system, public security system, other typical systems applied for typical civil and official system...

Because of the rising complexity and functions of modern systems (especially those large-scale and expensive ones), public service systems and other systems are prone to share or partially share infrastructures, and different systems are interlocked and intertwined, with more and more cross links.

For instance, GPS system of the US, which though originally only for military use is now put to civil use as well; on the one hand, it is of public service and national security type, on the other hand, it is of civil commercial type. Another example, military and civil communication systems share infrastructures of trunk communication networks.

Public service systems also don't take direct economic benefits into consideration, which only consider social interests permitted by national budgets and strive for more social benefits within the limits of national spending.

6.1.2.3 Economic Development Type (Economy in Broader Sense): Abundant and Common in Market Economy

Production is the most basic practical activities of mankind, which will never end. The development of production brings about the constant advancement and increasing complexity of productive means, exhibiting the progress of artificially designed productive systems, e.g., the manufacturing system of chips, which increases the performance of chips million-fold while reducing their prices at the same rate. Only in this way computers become so popularized and form networks. The development of productivity correspondingly pushes forward the expansion of exchanges, which by using markets as major medium promotes the rapid development of various artificially designed commercial systems, such as business management system of commercial banks, custom service system, communication system ... too numerous to enumerate.

6.1.2.4 Social Development Type

For instance, various systems applied to disaster prevention and reduction (meteorology, geology, oceanography, epidemics ...). Or environment monitoring and environmental protection, traffic and traffic management control, public health (first aids included), education, various social security systems …, which have become ones of the important indicators of the degree of regional and national developments. The more advanced the social development type artificial systems are, the higher the levels of regional and national development are, and the more complete the social development is.

6.1.2.5 Living Consumption Type

With the developments of science, economy and society, a large proportion of consumer goods are being vigorously developed from simple separate products to advanced, multifunctional ones that have characteristics of systems. The research and manufacture of consumer type artificial systems that arise out of this situation become ever more complicated.

Artificial systems of living consumption type are currently characterized by categorization and universal applicability in combination with personal preferences and needs. Though categorization (for instance, functions …) displays conspicuous multifunctional trend, whose interfusing and intertwining with other types of systems are intensified, and tends to blurring the "boundaries", as a branch of artificial systems, it on one hand is developing independently, on the other shows inclinations of merging with other types. Yet as living consumption system, satisfying the developing needs of people's living and high consumption must be the core factor of its survival and development.

6.1.2.6 Comprehensive (Mixed) Type Artificial Systems

Thanks to scientific, technological and social developments, the ranges of human activities are increasingly widening with their integration with nature more and more tightened and detailed. Therefore, artificial systems, which are in the service of mankind, must generally accommodate the various needs of the development of human beings, as exemplified in the dialectic development of the specialization of functions on one hand and the integration and diversification of functions on the other. Because of the progress of digitalized microelectronic integration and soft-and hardware modules, various functions of information systems are easily realized from the perspective of production, but perceived from the perspective of application, the integration of functions are not unconditionally supported, for both functional types will persist continuously in long-term confliction.

In countries, regions or societies where science, technology and economy are highly advanced, there are currently occurrences of the transfer of the objectives of some artificial systems. In the past, the transfer of the objective of an artificial system would give rise to massive chaos, or even lead to the failure of system design. In contrast, at present, in consequence of the diversification of the needs of social development, if one "objective" of an artificial design cannot be achieved (design failure), it can transfer its objective and metamorphose into another type of artificial system in order to partly retrieve the losses. The most notable case is the design of space shuttle by the US, which as a large-scale satellite launching commercial artificial system can't achieve its original objectives, but is successful after being changed into a large-scale scientific project! However, this kind of maneuvers is not very common. It depends on the specific conditions of the "functions" and "structures" of the system. For instance, "Iridium" communication system hasn't been able to be converted into artificial system of scientific engineering type, only to be sold to the military at a low price. Certainly, we are after successful designs that stick to their initial purposes.

Man and the related parts of cosmos that survive and evolve jointly with human beings work together to form a more massive survival system for mankind, which at the same time is the most intricate and inclusive "open complex giant system" that includes "numerous" artificial systems. Because the knowledge of mankind always lags behind the complex motion of objective things, and to make things worse there are the constraints of the practicing abilities of human beings and motivations of partial interests, many types of artificial systems are not in complete harmony, or even are incompatible with each other. Furthermore, every artificial system is accompanied by its own negative effects, in consequence of which it is far-removed from complete scientific order. On top of this, the severity of a few problems is gradually escalating. Therefore, as to whether or not human society, this complex open giant system, will perish in its own hands, disputes abound, with some arguing that the rationality of human beings will at least prevent them from destroying themselves, others put forward that they will probably be their own doom. The issue has acquired the status of a complex philosophical conundrum. However, in the design of artificial systems, it is of great necessity that more attention should be paid to the negative effects! We hope that human society will continuously evolve relying on the development of human reason and rationality.

6.2 Origins of the Development of Artificial System: Discussions on the Thinking Pattern of Mankind

6.2.1 The Essence of Mankind Lies in Their Learning and Practicing Abilities, Which Are Also the Fundamental Causes for the Development of Artificial Systems

The essential difference between human beings and other things (living organisms included) is the former's abilities of learning and practice under the guidance of reason, hence, their abilities of rationally and actively adapting to, integrating into and transforming nature. They are the common potentials of development, which are forever in application. It should be stressed that "learning" and "practice" are two different abilities, though interrelated and mutually developing, transforming and unifying. They are an important pair of the unity of opposites and in constant development and dynamic changes, whose law of development is the embodiment of the law of the unity of opposites in the rational evolutionary process of mankind.

6.2.2 The Core Elements for the Development of Mankind's Learning and Practical Abilities Are the Development of Thinking Abilities

It is mainly by relying on thinking abilities that in actual practices human beings see into environmental factors, and use the knowledge accumulated in thinking about solutions for problems in accordance with these factors. The specific contents and results of "thinking" are too plenteous for enumeration. However, since ancient times, people have been asking whether the combination of men's thinking and being (the most elevated "being" in the cosmos) has the most universal, general and truthful thinking pattern. This is the most essential problem of philosophy: What is "being"? A core enigma in the existence of mankind! Naturally if humans didn't exist, there would be no philosophical problems, while philosophy is crucial and indispensable for the understanding and improvement of man's thinking.

6.2.3 The Scientific Thinking Pattern of Mankind Should Be Dialectical and Follow the Law of Unity of Opposites

The law of dialectics can be simplified as: What the things in motion are and what they are not at once. This is because every objective being is motion, i.e., the dialectical motion of the unity of opposites. The existence of a thing is the synthesis of the thesis and antithesis, while its motion is forever conditionally towards the opposite side. The existences of human beings and thinking are the most complex and intricate motion of opposites currently known, which should best exemplify the laws of dialectical unity of opposites. After millennia of progresses and retrogressions, twists and turns, in combination with the development of human beings and society, mankind is only beginning to know the law of the unity of opposites, though this law is still now not universally acknowledged. I personally think the reasons behind this can be summarized as: Firstly, using "thinking" to know thinking is susceptible to the error of "I can't tell the true shape of Mount Lu because I myself am on this mountain"; secondly, civilization is still not very highly advanced, and only when it has reached advanced level, can the most essential basic law of complex motion be clearly understood. It is with the advanced evolution of mankind, including highly developed reason and rationality, the understanding of the laws of dialectics can be more and more deepened. From this, the conclusion can be drawn that: The knowledge of the law of the unity of opposites in constant combination with practices is the inevitable path of the development of mankind, the reflection of their existential mode (their initiative included), the methods to transform the world and the unification of their lawfulness and purposefulness.

Upon more detailed discussions, human beings have reason and rationality to contemplate on "why", "what", etc., which can be summarized as "what is being", and when themselves are related, they will meditate on "the existence of myself". However, these two problems can differ in fundamental ways, which should absolutely not be confused with each other (but easily confused). One is concerned with "universal" being, while the other is questioning a special being (Dasein). Descartes's famous dictum that "I think therefore I am" can be applied to the latter, but not the former. To know universal "being" (sein) through thinking is an extremely complex conundrum. If the concepts of sein and dasein are confused, the subjective and the objective, the individual and the universal are confounded as well, thus obstructing our comprehension of the law of the unity of opposites. Mankind should learn objective laws through thinking instead of replacing them with subjective thinking.

6.2.4 The Law of the Unity of Opposites is the Basic Law in the Research and Development of Artificial System

6.2.4.1 Artificial Systems Are Relation of Constraint Condition

Artificial systems especially those complex important ones in the process of scientific research and development directly require the support of system science, information science and control science to resolve the constraints and interrelations of complex structures, functions and conditions of multilayered sections that can be contained in systems. The formation of these relations is the concrete reflection of the objective motion of the unity of opposites in the process of the research and development of artificial systems, so the basis of scientific research is to correctly understand the essentials of the subjects by referring to scientific thinking, of which dialectical thinking is of great importance, and upon further analysis, the explorations and thoughts on the basic problems of artificial system design can all be incorporated into the philosophy of artificial system design, the core law of which is the law of the unity of opposites.

6.2.4.2 The Core of the Artificial System Is the Law of the Unity of Opposites

The formation, existence and decease of an artificial system are an unavoidable dynamic motion process, which is a complex motion process as well, while the most essential law guiding complex motion processes is the law of the unity of opposites. Therefore, in the representation and mastery of complex systematic motion processes, the supports of the disciplines of system, information, control and other related disciplines in combination with dialectical thinking pattern are indispensable for the successful research and development of complex information systems.

6.3 Philosophy of the Research and Development of Artificial System Design Under the Guidance of System Theory

6.3.1 From Philosophical Discussions to the Philosophy of Artificial System Design

6.3.1.1 Brief Summary of the Contents of Philosophical Researches

Human beings, the "crown of creation", possess inspirations and thoughts. Confronted with the existence of various things in the environment of their survival (including the being of their own existence), they naturally inquire: What is the "being" of things? Why do they exist in this way? That is, to search for the answers of the two basic questions "what" and "why", for which there are simply no direct replies that men can inclusively offer (especially in ancient times). However, driven by the mankind's knowledge-seeking instincts and their basic needs for survival, these questions are unswervingly and strenuously explored, for the most basic and universal deep-level laws, which is the mission of the philosophers that has lasted for two millennia. Due to the complexity of these problems, philosophy is deemed as the discipline where human beings endeavor to find hidden truth with reliance on their own wisdom. A famous Chinese philosopher concisely rendered "what is Being" into corresponding Chinese, and put it as the most basic proposition of philosophical researches, as illustrated in the following diagram (Figure 6-1.)

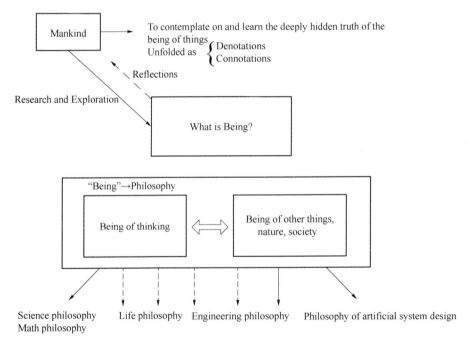

Figure 6-1   Content of Philosophical Thinking

The connotations consist of men's explorations of and reflections on the fundamental issues of themselves and the world they inhabit; the denotations are the formations of theorized worldviews on things at different levels guided by the spirit of seeking truth from facts and never-ending explorations.

Because the "being" of mankind (the "being" of thinking) is the most elevated form of complex "being", it is naturally included as subject of inquiries. The unique issue of applying thinking to study the "being" of thinking emerges, the complexity of which is demonstrated by dotted lines.

On account of the fact that the study of "being" is concerned with the universal basic laws of the motions of billions of things, these laws should encompass all the related "motions" to cover their essential and universally applicable rules, with merely the help of "thinking", which is also included as one of the research subjects (analogous to what occurs in mathematics when a set is its own subset, easily leading to paradox). Therefore, the laws of "being" in philosophy are extremely onerous to obtain, which after two millennia of mankind's strenuous efforts are still far from perfection, with predilection for one side or another that gives rise to corresponding schools or factions. All those contentions and controversies generally can be attributed to two cases. The first is that of the nonexistence of the system of research subjects, as Ji Xianlin perceptively pointed out: "No matter what religion in the world, when studied with scientific methods, will turn out that its doctrines and rituals all have historical processes of development and origins of coming-into-being, and they are all made by men, full of contradictions and flaws ..."The second is that due to the complexity of the problem, people haven't been able to grasp its essentials, so consequently, everyone has his or her understanding. Since the motion of the cosmos is objectively existent, the first reason can't stand. Thus only the second case can be held accountable for the arguments concerning the laws of "being". I am personally certain that the most basic law of "being" is the law of the unity of opposites as proposed in materialist dialectics, though its explanation of "being" in combination with realities, its guidance of present "being" and its development of future being are still not full-fledged! Besides reflecting on "being" in general sense, humans often seek basic universal laws of individual fields combining their own practices, which forms diverse philosophies concentrating on different areas. We are now discussing the issues of artificial system design, and in order to strengthen the groundwork, it is preferable to deliberate on the philosophy of artificial system design, from the two perspectives (its connotation and content) that consist of four problems:

(1) The connotation of artificial system design;

(2) The laws of its existence (survival);

(3) The basics of the existence of its designers;

(4) The formation of design criteria by the combination of designers and the system of the subjects of design.

6.3.1.2 Brief Introduction of the Connotation of the Philosophy of Artificial System Design

- As a branch of philosophy, the philosophy of artificial system is also a discipline that has affections for wisdom and devotes itself to the pursuit of knowledge and hidden truth, only that this field of "learning" is limited to the related areas of artificial system design. However, the development of the philosophy of artificial system design will also enrich the general philosophy.
- The basis of the philosophy of artificial system design should be the law of the unity of opposites of the motion of things and modern system theory (As the universal theory of the survival and development of all systems, it should largely correspond with above mentioned philosophical laws.), which jointly constitute the groundwork of the philosophy of artificial system design—a two-basis theory.
- The philosophy of artificial system design studies two domains of "being", which are not the common being as concerned in philosophy, but the being (social being) of artificial system and the "design being" of men. The latter means the integration of the laws that should be followed in "design" into artificial systems, making them comply with the laws.
- The functions of the philosophy of artificial system design in practices are to boost the constant development of new tools indispensable for the evolution of human society through the combination of science and technology with philosophy, forming an important content of the self-organizing social progress.
- Corresponding to the reflections and studies on the fundamentals of self and human society in philosophy, the connotation of the philosophy of artificial system is included in the basic issues covered by the two domains of "being", which is definitely related to human reason, common sense as well as physical and physiological problems. It should be said that it is an unavoidable rule, which emblemizes the long-term efforts of learning and practicing that embodies the evolutionary process of human beings.

6.3.2 Brief Introduction to the Philosophy of Artificial System Design

6.3.2.1 Basic Laws of "Being" of Artificial System

- All artificial systems are mainly systems controlled by human beings to serve their needs. Therefore the self-organizing functions of systems are designed by men and exhibited in systems, running in self-organizing manner and delivering "services". In circumstances of social changes, and the needs for services to decline or perish, artificial systems will demise too. Thus we can conclude that the lifespans of all artificial systems are finite, or even comparatively fleeting! — the law of the finiteness of life.

➢ Modern artificial systems generally exhibit the concentration of service functions, while the formation of specific services is through the self-organizing mechanism of systems in combination with the application environment, forming operative procedures to provide diverse services. Therefore, it is important to study the dynamic living self-organization mechanism as a foundation.

➢ Though the lifespan of artificial system is limited, trying to extend it is the common desire of all related parties. In its life of service, service functions will undergo the development of quantitative to partial qualitative changes(The theory and method of mult living agents in artificial systems is one way to enhance the service quality of artificial systems and extend their lifespan.)

➢ The specific causes for the demise of an artificial system:

√ Social changes as well as the changes and demise of the mother system in which the artificial system is contained result in the degradation and decline of the service functions of artificial system.

√ Social changes as well as the changes of the mother system lead to the changes of constraints, which the service functions of the system can't meet, and therefore cause the demise of the system.

√ Social, scientific and technological progresses will give rise to similarities between service functions, and consequently new systems of fewer costs replace the previous systems.

√ When there is a substantial increase of the negative effects that exist in the system, if certain limits are surpassed, this artificial system must inevitably die.

√ When the direct service functions of the systems disappear, a few artificial systems will change into cultural objects, playing cultural functions and continuing to survive for a long time.

➢ When one kind of artificial systems deceases, many demands for the birth of new artificial systems will necessarily arise. Supported by science, modern society is developing more and more rapidly, and the depth and ranges of mankind's activities are also substantially increased. The unilateral satisfaction of human desires has led to unsustainable development as well as disharmony between men and the environment. Reason helps us increasingly perceive the severity of this issue, and from now on, man will strengthen the transformation of the survival and development mechanism of artificial systems: from the mechanism of boosting the development of artificial systems merely through considering service needs, to that of comprehensively considering "needs" and the environmental factors of sustainable development. The deep changes of social knowledge will also be embodied in the survival mechanism of artificial system—the development of survival and development mechanism.

6.3.2.2 Basic Laws of the Success of Design of the "Being" of Men in Design

> Mankind is the subject of the birth of artificial system which originates from the "objectives" of mankind's needs for services. It is a kind of "future", which is specified by "objectives" into "needs" and the planning and determining of the satisfaction of "conditions", and then objectified into the realization of artificial systems. This whole process is artificial system design. In the actual survival of mankind, numerous "needs" arise that should be addressed, which is naturally indubitable. The key is to first transform needs into "objectives" and desires under certain constraints (the two must be coalesced), the process of which calls for the combinations of imaginal thinking and logical thinking, and quantitative and qualitative changes, while considerations of constraints shouldn't be restricted to all direct conditions, but should include all necessary environmental conditions and changes of constraints contained in the changes of these conditions (deep-level change factors) as well. The process of this transformation is targeted at truths hidden behind the facades of things. In the planning and designing of forming "objectives" and "needs" of large-scale important artificial systems, core "objectives" and "needs" should be emphasized, combining "human reason, common sense, physics and physiology". It is of great importance to avoid going against the "truth," which results in failures.

> The "being" of artificial system design is a kind of exchange between "acquisition" and "expense" under certain conditions. In order to acquire something in the aspect of service functions, some "expenses" have to be paid (increasing constraints included). Certainly, "acquisition" can be in the domains

of conditions, while "expenses" in the domains of functions, form exchanges. "Exchange" is formed based on "human reason, common sense, physics and physiology". In the application of these rules for above-mentioned exchanges, primary service functions should be accentuated. There shouldn't be too many other functions, in which case "expenses" will be increased, surpassing limits and leading to failures. Whether or not there are contradiction and incompatibility between "acquisitions" (service functions) and "expenses" should be considered as well, the solution of which is the second problem of the unity of opposites that should be taken into account in the process of design and planning.

➢ For any "acquisition" in the artificial system, there is corresponding "negative effect", and "positive and negative" effects are exhibited by the transformation of conditions. In the design of artificial system, the designers should thoroughly deliberate on the issue of the occurrence of negative effects, preventing that they are generated by services in application because of changed conditions, which further influences the existence of the system. This is the third problem of the unity of opposites that needs to be dealt with.

➢ The "being" of artificial system to a large extent depends upon its self-organizing mechanism which is indispensable for the smooth fulfillment of normal services in the process of its coming-into-being. Preventing accidents and matching with the environment have become the main mechanism of the survival mechanism. Because of the complexity of present artificial systems and their application environments, it is important that in the implementation of the requirements mentioned above, emphasis should be put on applied computing sciences (model-building and simulating computation for the analysis and solution of problems, which is then reflexively mapped to the practical space, is the latest efficient method).

➢ Besides considerations and guarantees for connotations of science and technology in the being of artificial system, the harmony of its different parts is another deep-level factor that should be incorporated into the design process.

6.3.2.3 Guidelines that Should Be Followed in Artificial System Design (Rules that Must Be Adhered to Deduced from the Successful Mechanism of Design)

The following guidelines shouldn't be disobeyed in whatever situations in the process of design. It often occurs nevertheless that even if men appear to be doing the most natural thing, they are already breaking the "norms", leading to their retributions.

- ➢ The research, development and design of an artificial system should be based on an obtainable objective, which shouldn't be altered. In the case of deviation from this objective, a whole new mission is to be started anew.
- ➢ In the design of artificial systems, as to applicable tools, we shouldn't merely pursue innovations and advancements. Under the condition of thoroughly meeting the "prerequisites", the simpler and more matured the tools are the greater chances of success they will have. The law of the unity of opposites applies between the level of the development of system and the related support of technology.
- ➢ In the design process the focus shouldn't be solely on the requirements of individual indicators. It should also take into consideration the core functions in the running of the entire system corresponding to the requirements of the procedures of the operations of self-organizing functions. For instance, when a number of important secondary indicators have reached the "requirements", but the margin is not significant, the system as a whole won't function well (Technical requirements of artificial systems are distributed from the top to the bottom, while in the design process comprehensive assessments should be pursued in the opposite direction.)
- ➢ As to a complex artificial system, in the process of its design, the scientific integration work of the whole system should be taken notice of, for only through correct systematic integration, can works that are divided into various components be united into an entirety, i.e., to realize the qualitative change that "the whole is greater than the components added together". Particular attention should be paid to the incorporation of the "integration" work into the whole design process.
- ➢ In the process of the design of important artificial systems, the dynamic changes of the constraints should be noted, as well as the incompleteness and non-self-consistence of the deep-level denotations of those constraints. Related measures should be taken to enhance the designs' abilities of adaptation, in order to avoid total failures.
- ➢ The lifespan of an artificial design is not infinite. As a consequence, how to scientifically deal with the decease of artificial design is a complex issue and a challenging problem, for in addition to coping with its demise, issues related to

larger scopes are concerned with, e.g., after ending production, how to provide users long-term service support, etc.

The overall principle is to scientifically and lawfully deal with the decease of artificial system, not to go against the laws of the survival and demise of systems, elongate by force the irreversible expiring process of the self-organizing functions, that is, the life of a perished artificial system, or abnormally shorten its lifespan. Generally speaking, the demise of an artificial system should be a declining process instead of a violent death, in order to better cope with the follow-up or replacement handlings of its service functions.

Follow-up handlings include those of the ending and transformation of the services, while replacement handlings include those of the assumption of responsibilities by substitute or new systems, etc.; "death" means the end for the deceased system, but "new birth" for the following system. We say, "if the old doesn't go, the new won't come".

In the early stage of the design of an artificial system, if there is in this early period of development possibility of adjustments for its structure, there is possibility for its improvements and enhancements, to which due attention should be paid.

Though the above discussed artificial system may be complex, both its main "requirements" and restrictive conditions are clearly depicted. There is yet a kind of complex systems which are hardly in congruity with the above illustrations, only to be contemplated on by referring to these thinking patterns mentioned above, instead of in the manner of completely quantitative thinking.

6.3.2.4 Comprehensive Case Studies of the Application of Design Philosophy

Case 1:

Around 1910, Ford designed Ford Model T, so that the number of car ownership developed from the ratio of one car in 150 million Americans to one car in hundreds or dozens of Americans. A total of 1,500 million vehicles were produced (In 1927, when models were not so updated as now on yearly basis.), which was a huge success of design. The core of its design was to facilitate mass productions, so that ordinary workers could afford and were willing to buy cars, thus dramatically boosting the development of the enterprise.

Case 2:

Cases of failure often occur due to excessive enlargement of secondary functions, e. g., the setting up of a small screen of a different channel in the screen of simulation television. Because the formation of multiple screens in the simulation system brings about massive costs, and the rate of performance and price is disproportionately low, those televisions with auxiliary screens widely advertised at that time haven't survived in the market.

Case 3:

The optical communications products of the American company Lucent had the advantage of quality and performance, but were put to the market six months later, which allowed their competitors to take more than half share of the market. To make things worse, they had overestimated the demands for optical communications products, resulting in the languishing demands for high-end products of more than 40G. This led to huge losses to the department of optical communications, which had to be closed, and eventually to a large extent to the selling and reorganization of Lucent Technologies (including Bell Labs).

Case 4:

The Failed Maiden Launch of Ariane 5 as a Result of Incorrect Integration

Ariane 4 was a very successfully designed rocket, which had been successfully applied for commercial satellite launches many times (including geostationary orbit). Based on this model, Ariane 5 was designed with more effective payloads. Because of the basis of software designing and spatial data fully accumulated in the past, the chief designer and the manufacturer decided that the first launch should be a commercial launch, which moreover was not insured, while satellite manufacturers and space experts of the world were invited to witness the launching ceremony. Propagation was strengthened to enhance competitiveness.

The consequence was that only 40 seconds after launch, at the altitude of 4000 meters, it deviated from the previously designated orbit, which couldn't be corrected. It ended in explosion.

Key Points of Rocket Flight Control:

Rocket control and the acceleration of space velocity are implemented by the inertial measurement system (including the laser gyro, accelerometer, and computers).

The control of rocket flight system or the implementation of flight is executed by the rocket computer.

In the launch phase, the inertial measurement system in 60 seconds after the launch implements north-positioning program, and after 60 seconds implements attitude control, while speed measurement and flight control system jointly control rocket flight.

Direct causes for that launch failure:

After about 40 seconds (not necessarily 60 seconds), Arione 5 began the angular rotation, and the inertial platform computer mistook computer data for abnormal (deeming errors spilling over), immediately turned to the backup computer, which spilled over too. At this moment, inertial platform computer was not equipped with flight program data, and in consequence could neither test whether the flight was normal or not, and nor send data to rocket computing programs. In this case, simplistic shutdown was entailed, and thus the rocket computer couldn't correctly control the flight, leading the rocket to deviate from the orbit and explode.

Causes for launch failure of the entire system-incorrect integration of the system

The initial launch program continued to use the 60-second period of Ariane 4 to replace 45-second period.

Fault-tolerant design was erroneous, and double warm backup only applied for accidental random errors, not for systematic program errors; it was not equipped with second-level error-sensing design, and the inertial platform computer was not equipped with flight path data (No method was taken to deal with the problem of insufficient capacities.);

The measures in system design were not comprehensive, which were one-sided and inconsiderate, e. g., continuing to apply Ariane 4 launch software for fear of the unreliability of new designed software, which on the contrary only led to incompatibility of the software with Ariane 5;

In the simulation experiment, close-loop simulation of the whole system was not carried out, and the actual difficulty was that linear acceleration of very high-speed was hard to simulate;

In the segmented subsystem simulation, data input was incorrect, far from accurate. Therefore, though simulation experiment was carried out with high marks, it didn't have practical significance.

To conclude, incorrect integration misled directors in charge of design and development in many simulation experiments, precipitated them into sudden failures, when everything went deceivingly well, causing catastrophic losses.

Case 5:

Edge design of "the affirmative as well as the negative" will play an excellent role in certain occasions.

In War II, Germany's mini-battleships had higher speed than ordinary battleships, heavier-caliber shipboard artillery and thicker armors than the cruisers, with great performances of autonomous operations. They were used to cut off very important traffic lines of the allied forces (besides the application of submarines). When confronted with cruisers and their formations, they would attack and annihilate them (especially those important transportation fleets). If encountering formations of battleships, they would attack their cruisers and then immediately run. They were very difficult to stop without satellite reconnaissance. The British took huge prices to finally wound them and let them sink blocked in the neutral ports.

The America-designed P2-V7 low altitude reconnaissance aircraft was equipped with propeller engine as well as jet engine (auxiliary), designed with great low altitude performance and flexibility. In combat missions at that time, when confronted with a jet fighter, it would rely on its flexibility to escape its attack, and when confronted with a propeller fighter, it would depend on its jet engine for immediate escape. Thus it was very successful at the time. However, with over-the-horizon missiles, a jet fighter can now strike it down without close engagements. In consequence, the design of P2-V7 is outdated now.

Case 6:

Sanmenxia Reservoir of China was designed in accordance with the comprehensively formulated design plan of some archetypal reservoir yet with enlarged capacities. Its basic design was wrong and the construction of the project failed due to inadequate consideration of the high sand proportion of this reservoir (one of the main constraints) and erroneous assessment of the influences of sands.

Case 7:

The development of manned spacecraft in China which adopted the form of spaceship instead of space shuttle exemplifies a successful philosophical practice in design. In the development of manned spacecraft since the mid-1950s, America first successfully invented the space shuttle, and other countries such as Britain, France, Russia and Japan followed suit. Based on comprehensive debates about the overall plan of China's development of manned spacecraft, conclusions were drawn that in the early phase of the men's entry into the outer space, the missions of manned spacecraft are not yet very frequent and imminent, while scientific and technological basis for frequent uses of manned spacecraft is not yet sufficient, which brings about huge costs and risks. Therefore, in addition to enormous expenses and hazards, the needs for main functions are not yet adequate, leading to the evaluation that the plan for space shuttle is not yet mature and practical. The designer in chief of manned spacecraft of China agreed with the above scientific conclusions, vetoed the space shuttle plan, and chose the spaceship instead that could only be used for once, thus smoothly carrying out the early plans of the development of China's manned spacecraft.

Chapter 7
Major Points in the Design, Thinking, Methods and Process of Artificial Systems

7.1 Preface to the Significance of the Expansion of Artificial Systems

7.1.1 Major Artificial Systems Are Already an Indispensable Part of Human Society, Whereas an "Innovative Society" is the Scientific Aim and the Core Representation of China's Social Development

Artificial systems are now an important composition for modern society's existence and development. Though lifeless, they are indispensable to individuals and human society as a whole, some of which are considered as inseparable parts of human body. And this trend is still expanding. Some artificial systems (or artificial manufactures) which once played a major part in history still exist often as human civilization (specifically as part of history), and are still an important constituent of human society. Besides, the state of development of major artificial systems symbolizes the degree of social advancement and the driving forces for development, and is considered as a key factor in inter-society competitions and struggles, of which both our authority and people have a sound understanding and have accordingly summarized and promoted it to a higher degree as "constructing an innovative society" which fully illustrates the long-cherished wish as well as principle that society develops with high quality and at high speed, in which advanced artificial systems continually spring up, to eventually realize the harmonious co-development of both small and large cosmos.

7.1.2 The Birth of Fine Artificial Systems Involves the Application of Various "Law" Systems, and the Birth of More Fine Artificial Systems and Their Being Incorporated Into Society to Function Are Strongly Indicative of an Innovative Society

Important as they are, the birth of complex major artificial systems is a rather arduous process, which starts from basic scientific researches, going through a long procedure from the promotion of relevant technologies, the system research and development and design, production and manufacture, to the state of "common development" of their universal and effective application in society. The principle (law) of artificial systems' birth and survival concerns basics of philosophy, and extends to the four domains: laws about humans, laws of things, laws of affairs and laws of beings, which interweave and interlink mutually and need overall planning and all-round consideration. Furthermore, the incorporation of artificial systems into society has to be within bounds, the birth and application of major fine artificial systems are therefore a great concern and need seeing their implementation in the sustainable development of an innovative society.

7.1.3 Innovative Society—the Core Factor in the Harmonious Development of Man and Society is that Human Aptitudes Continually Develop Under the "People-Oriented" Principle

Innovative society is the fine state in which man and society achieve harmonious and sustainable co-development, and its core factor is that under the "people-orientated" principle human aptitudes continually improve along with the steadily advancing human living conditions (including environmental conditions) and quality. And human aptitudes' continual development is based on the development of educational system, specifically, the reform and development of the complete educational system starting from preprimary education, through elementary and secondary education to higher education. A modernized educational system aiming at improving education quality is therefore important. The development of education must have strategic priority in society, and the disciplines of science, engineering, management and economy in higher education are the immediate base of the research, design, manufacture, operation and management of artificial systems.

7.1.4 Thinking and Thinking Methods

Human thinking ability is the core of human cognitive and practical abilities, and the most precious ability unique to humans only among all things on earth. But "thinking" is also the most complicated and finest motion in the world. In the evolution of human thinking, the inner (inside the brain) fine motion is the production of a long process of evolution, which can combine and utilize outer motions to serve humans. Language and writing system resulted from the evolution of brain and thinking are the condensed reflection (in constant dynamic changes) of human cognition of the outer world, and the tool for human communication and cooperation in certain scope. Human thinking and language and writing system enable human beings to continue and promote at once human cultures (civilizations).

Human beings' possessing thinking and cognitive abilities naturally render them eager to understand the essence of human thinking. Actually man has been working hard in this field, yet since thinking ability is of the highest level among all abilities, man's efforts to understand thinking are eventually to understand "self" via "self". In the field of cognition, there is nothing more advanced than "thinking" to help humans understand "thinking". Consequently, the degree of understanding "thinking" remains uncertain, and has even evolved into a philosophical proposition (As a classical Chinese poem goes, "Of the mountain Lu we cannot make out the true face, for we are lost in the heart of the very place", which means we fail to see the truth about a person or a matter because we ourselves are obsessed in it.) But cultures formed in this process, especially the spiritual culture, are of great help to and enlighten each one of us. "Thinking" is generally taken as human beings' most important ability; besides this ability (vital to individuals), there are other human "achievements", including personal life philosophy, ethics and morality, and etc, which are actually a reflection of "law about humans". The combination of human thinking ability in general and the thinking "achievements" by individuals further affects the extension of the process of human thinking motion. Man's long efforts of auto-gnosis of the thinking principle and methodology have also achieved positive results including types of thinking, thinking methods to understand things (including "analytical" and "synthetic" methods, and methods using the correlation of things to understand new things, e.g. analogy, association and transformation). Yet things in the world are in ceaseless motions, having countless categories, properties and principles; classification and division of thinking methods resulted from summarization and induction are not complete, and are in the process of constant development and completion. But certainly the positive results achieved on thinking methods are beneficial and worthy of popularization, communication and discussion. Therefore, basic thinking methods will be discussed in the following sections.

7.2 Types of Human Thinking

7.2.1 Logical Thinking

Logical thinking is that humans use concept, judgment and reasoning, with the help of language representation, to do thinking operation in hopes of grasping the essence and motion principle of things. To do right logical thinking, one has to know well two essential factors among others: subjective consciousness and objective material factor. Subjective consciousness is the initiative subjective factor which reflects the people-orientated principle, which means that when conducting right logical thinking, humans have to select from their large knowledge theoretical system the suitable serial ones to judge and reason starting from conception (Certainly this has to combine with objectivity, i.e. the material factors to be discussed.) This proves that human subjective consciousness is indispensable. The second essential factor, the material factor of object's objective existence, accentuates that in operating logical thinking, humans have to admit and respect as well the objectivity of the object's material motion. Humans need to extract the essential principle of the object's motion via the information produced by the motion to complete right logical thinking, and eventually get the correct answer. So right logical thinking comes from the correct combination of the two "essential factors".

7.2.2 Figurative Thinking

Figurative thinking is a complicated thinking, and its mechanism remains unknown, at least partially. It differs from logical thinking in that it doesn't use conception, judgment and reasoning but imagination and symbolization of related feelings. It is operated via images and models of images rather than language presentation, and the image formed by which is of advanced properties in that such an image is not the direct reflection of the "object's" image, but the idealized image of the development of the future object (the object of thinking) which doesn't yet exist in reality and can't be seen of the substance (a reflection of "being without form" in Taodejing by Lao Tze in practice). The result of figurative thinking which is of different category from logical thinking is a thinking creation, a successful one, and can reflect the essential principle of the "object".

7.2.3 Intuitive Thinking—Function of Human Consciousness Which is of Deeper Level than Figurative Thinking

Intuitive thinking is to maneuver the thinking achievements buried deep in sub-consciousness, especially that of figurative thinking, and the achievements of psychological experience gained via appraisal and appreciation of the beauty (Note: the connotation of the word "achievement" here implies that it's not gained out of the void, but the valuable results stored in human brain accumulated through previous thinking activities.). This type of thinking, communicating with the problems to be solved in dominant consciousness to achieve a burst and insight solution to the problem, doesn't follow the normal thinking process from the perceptual to the rational, but goes like this: After repeated efforts, still no solution is found to the problem and things seem to reach a deadlock. Keep trying and never give up, and suddenly get enlightened with the help of accidental opportunity, then core methods were found to solve the problem. This is a leap achieved with the accompaniment of imagination and figurative thinking. But to achieve a comprehensive solution, the result of intuitive thinking needs to be completed via logical thinking and accurately and precisely expressed in written language and eventually put into practice. Intuitive thinking doesn't come from intuitive "sub-consciousness", nor illusory groundless "sense of gods", but is a leap thinking qualitative change resulted from the combination of sub-consciousness and the dominant consciousness intending to solve the problem in human brain.

In practice, the solution to comparatively difficult problems needs repeatedly using these three types of thinking and their combination as well.

7.3 The Transcendental Promotive Thinking Mode (Combination of Methods and Simulation): Following and Using Philosophy to First Impel the Transformation of the Conditions for the Development of Things, Then Impel the Divergence of the Scientific Evolution of Things

7.3.1 Philosophical Basis

Dialectical materialist philosophy holds that all things exist in contradictory motions, the specific existence of a thing is a transient unity of contradictions under certain conditions, the general trend of the motion of things is the unity motion which negates reality, and the evolution of specific things is realized based on specific conditions. In Chinese philosophy, e.g., great thinkers represented by Lao Zi use the idea of "being with form" and "being without form" to accentuate and depict the philosophy that things transcendentally develop following the law of the unity of opposites. Everything on the earth is generated by "being with form", and "being with form" comes from "being without form"; The Tao in its regular course does nothing (for the sake of doing it), and so there is nothing which it does not do. Happiness is to be found by misery's side, whereas misery lurks beneath happiness; the motion of Tao by contraries proceeds, etc. Materialist dialectics and the transcendental and promotive of development and change philosophic thoughts in Chinese philosophy are the philosophic basis of human divergent thinking mode which is transcendental and promotive of development in reality.

7.3.2 The Transcendental Promotive Thinking Mode, the Core Factor and Move of Practical Application

A. Get familiar with the domain properties of the object; use logical thinking, figurative thinking and the combination of the two to understand the present state of motion of the object, and then find out and ascertain the direction of development from the directions of the negative phenomenon, including the development direction of the qualitative change with negative factors inside (not the direction of retrogression).

B. Deliberate upon the conditions necessary for understanding the development and change of things, and the conditions for "retrogression" as well.

C. Deliberate upon the methods of generating conditions for development and evolution (or the condition generating conditions for development); the generation of methods, especially good and important one, is an innovation of mental work, including the factor, methods and functions of beauty, and the factor of "ingenious" innovation, which eventually form "wonderfulness" (both beautiful and ingenious). The innovation of wonderful "methods" comes from the subjective factor, i.e., mastery of the characteristics and principles of contradictory motion taking place in the domain of the object (formation of knowledge), and the correct and clever handling of all types of thinking, and the combination with objective factor, i.e., respect the specific motion principle and state of things. Correct combination between the objective and subjective factors is thus formed.

D. In practice, the above steps need repeated adjustment and revision to achieve the combination between thinking methods and practice.

7.3.3 Exemplification

Cases of application:

The construction and development of social medical system evolves with the development of social science and technology and economy from confronting diseases and sickness to ensuring people's health—its corresponding aim is "no disease and sickness", which is "being without form" and transcends all diseases and sickness.

In cultivating university students great importance is attached to the cultivation of quality and innovative abilities, which is a kind of cultivation transcending the mastery of all specific knowledge and abilities.

During the anti-Japanese war starting from the year of 1937, China by adopting a strategy suitable for the protracted war deliberately abandoned some large cities and part of the land as well after battles, which confused the Japanese and presented them a false picture of victory. Under the false belief of their victory, the Japanese army had to scattered its armed forces to cover those cities and land. Besides, being unjust, it lacked popular support. Under this circumstance, its military strength was continuously consumed and lost all its military superiority and in the end was defeated by China.

7.4 Thinking Methodology for the Solution of Problems

7.4.1 The Direct Method Conforming to the Consequent Thinking: Grasp the Sensitive Principal Contradictions (in the Current Phase), and Take Measures Following the Direction of Development to Solve the Contradictions Hindering the Development of the Principal Aspects

This method is familiar to and frequently taken by people. There are two points need to be emphasized here: One is that from the representations of numerous phenomena people need to accurately pick out the principal contradictions (only in a small number, usually one or two) and their principal aspects to take effective measures to impel the secondary contradictions representing the new motion mode to transform into the principal aspects of the contradictions; the other is that never ignore this universally existent and applied principle—the "principal contradictions" are in a minority, and the two sides are antagonistic to and combat each other. This is a process in the dynamic development and need continuous development to hold the principal aspects of the contradictions.

Case 1: Resection in Western medicine. This operation's original starting point is that since the harmful doesn't disappear by itself and is "unnecessary" human tissue as well, the solution is to remove this harmful unnecessary via operation, which is of course feasible but with the most essential constraint condition of being bearable to people, and endeavors for the condition of people "losing" less and "suffering" less which eventually evolves into the surgical methods applied to diseases' early discovery and diagnosis and supported by advanced science and technology.

One case in point here is the surgical resection of tumor.

Another case to prove the above is the surgical removal of cholelithiasis. Originally laparotomy, this surgery, effective as it was, encountered the problem of patients' long cut and recovery time. The contradiction therefore evolves into the solution to the problem, and consequently this surgery is now simplified, in some cases, into one that need no long cuts and suture but three small apertures with the help of laparoscope.

Case 2: The reason why China can feed 20% of the world population with only 7% of the world land, advancing towards a well-to-do and even moderately developed society largely lies in its basic guiding principle in agriculture which accentuates "high yield" and "expansion of sown area", e.g., hybrid varieties, diversified seasonal varieties and methods, expanding the sown area, and controlling production, exchange and consumption to ensure their ordered operation. All these methods have achieved phasic effective results, but are now turning from merely stressing on the increase in quantity to the structural improvement of high yield stressing on the increase in quality. This is a case in point complying with the direction of development.

7.4.2 Reasoning Complying with the Direction of Development +Jump Solutions (Reasoning Following the Forward Direction in General Level + Jump Development in Key Technical Level)

The obvious improvement and notable leap in technical level are actually important technical innovations. But leaps of this kind can never take place out of the void.

Chapter 7 Major Points in the Design, Thinking, Methods and Process of Artificial Systems

Discoveries often appear in the form of insight, but are actually inevitable accidental phenomena. When the "base" is bred to certain degree to incur sudden changes in "practice", discoveries inevitably take place, though the specific time, place and the subject who proposes these discoveries are often random. Furthermore, under naturally mature conditions, there may appear the phenomenon that many people simultaneously bring forward the same proposition. Applying technical innovations to the key points in the design of artificial systems is an effective method in the research and development of new systems, but need weighing its advantages and disadvantages to decide whether it's feasible in general.

Case 1: the development of DSP. Information signal processing often uses some algorithms (which are also "transformations", e.g., Fourier transformation, Hilbert transformation, and etc.) which were first operated on computer, and later to save the operation time, FFT was proposed. This, however, was found not fast enough in speed and there were difficulties in computer programming. A leap thus emerges in the process of development, i.e., the creation and development of DSP which is a transformation and leap in the design of CPU and based on the structural system of computer. DSP continues to today and is still in constant development as a chip sequence.

Case 2: the invention of telephone. The invention of telephone involves the development of routes realizing direct conversational communication based on the electric communication (Telegraph was invented first which first transfers information to scripts, then to telegraph code to eventually realize the transmission of information.) The essential point in this invention is modulating successive speech sounds to electric signals to enable the latter to represent the former. This is a leap (But now the successive signals are discretized to form digital signals.)

7.4.3 The Opposite Yet Complementary Methods Using the Reverse Thinking to Solve Contradictions in the Light of Specific Problems, in Which the Important Thing is that the Opposite yet Complementary Effects Can Eventually be Achieved

The essentials of a forward development often bring about a series of new relations, while the so-called reverse thinking refers to the thinking mode and method opposite to the traditional way of solving contradictions. In fact the existence of "contradiction" corresponds with the existence of numerous "relations", e.g., if a method can diminish one of the necessary and sufficient conditions (relations) for the existence of certain "contradiction", it equals eliminating this "contradiction". This method is generally opposite to the traditional solution which destroys certain condition (Instead, this method may help the development of this certain condition.), and therefore called reverse thinking method.

Case 1: The functions of fighter planes, already very complicated, now stress on launching attacks before the enemy out of the range of fire to reduce losses and get upper hand. With the antagonistic development of both sides, fighter planes are even more complicated, expensive and heavier. In this process, new technologies are constantly applied to solving contradictions, but a series of basic contradictions resulted from the larger bulk, heavier weight (Fighters have reached a weight of 40 tons.) and higher cost are becoming more serious instead of being solved. A new method emerges which abolishes man pilot and reduces comprehensive functions (or moved behind for second-front support) to cut down the weight, bulk and cost. This avoids causalities of man pilot even when shot down for one thing, and the possibility of the fighter's being shot down also decreases with smaller bulk, lighter weight, and consequently improved mechanical performance and lower target reflection and radiation degree. Unmanned fighter planes are now being applied in practice.

Case 2: The thinking mode of Cao Cao (a personage in the period of Three Kingdoms) is different from the ordinary and he is therefore regarded as "oversensitive". Yet in fact his way of thinking is thinking subtly over the other side's thinking, which is a reverse thinking mode in a broad sense. Cao Cao is expert in multiple and diversified reverse thinking mode to achieve the "opposite yet complementary" effects. While thinking the other side's thinking, he goes a step further to calculate the other side's thinking methods to be used to oppose him, based on which he forms his methods opposing the other side's thinking methods used to oppose him. This is actually a "reverse reverse-thinking" method, which, in general, a more complete dialectical reverse thinking method and more advanced than ordinary ones. But this is not of the absolute. According to the chapter telling the story about Huarongdao in Romance of Three Kingdoms, Kong Ming captured Cao in Huarongdao by acting in a diametrically opposite way responding to Cao's reverse reverse-thinking mode. Cao knew that Kong thought him "oversensitive", i.e., doubting there were troops in ambush in Huarongdao and naturally not taking that way, and consequently surmised that Kong would accordingly not station troops there. He therefore concluded that taking Huarongdao was safe. Yet Kong inferred from Cao's usual way of thinking (reverse reverse-thinking mode) to make a correct judgment: stationing troops led by Guan Yu there, a dialectical thinking method responding to Cao's reverse reverse-thinking mode.

7.4.4 Analogy and Association

Analogy is to surmise from similarity in known respects to similarity in other respects. One case in point here is that legend has it that Gongshu Ban invented saw because of being inspired by jugged grass able to cut people's shanks. The conclusion from analogy needs of course further verification. Association is a thinking mode of linking certain concept to related concepts, i.e., a process from associative factors (the cause) via the associative route (knowledge + imagination) to achieve the associative effect (the combination of discoveries). Association is a "discovery" method freer, more active and divergent than analogy (which stresses on and generates based on the alikeness and similarity between two respects in nature). Induction is another "discovery" method. These three differ yet share similarity in terms of "discovery". In the process of thinking, the combination of the three will achieve a better effect. But the three methods function differently from deduction in that the latter is the reorganization of found "truth" while the former ones are to discover "truth". System designers attach more importance to the discovery of "truth".

7.4.5 Transformation

Transformation is a process by which one figure, expression, or function is converted into another one of similar value, e.g., one problem is converted into a known one (or its combination), or a simpler one (or its combination). As a universal method applied to many fields, this method can be used in different stages in the design of artificial systems, including the stage of the formation of system scheme, the stage of specific technical design, and etc. In system design, the most vital thing is to integrate the principles of system theory with the transformation method to make both more complete. For example, "function" and "structure" work dialectically to form an artificial system, in this process, we can transform the function and the structure (partial and whole); in searching for "order parameters" and arguments, transformation can be applied while in modeling we can also use this method. Find the intrinsic factors out of special problems to induce them into common problems, whereas grasp the intrinsic factors out of common problems to solve specific special problem—these different approaches actually contribute to the same end. The transformation method can be applied in the process of modeling, and the relation mapping inversion method as well, vice versa (including the analogy and association methods discussed above).

7.4.6 Basic Thinking Methodology Used to Solve Specific Problems: Combination of Qualitative and Quantitative Methods (Also Taken as One of the Thinking Modes Solving Specific Problems)

Qualitative and quantitative thinking are two basic human thinking modes (forms), the combination of the two composes human beings' specific thinking, and lack of any one or inappropriate combination of the two will lead to the failure of composing scientific thinking. The importance of their combination lies in the complexity of things' motions (system motion) which are mutually related in multiple levels and sections and therefore need both in-depth quantized researches and general mastery of the motion properties and principles as a whole. Only with meticulous quantitative researches and general researches of the overall situation can the complicated motions of things be mastered. Following the theoretical understanding of the importance of combining qualitative and quantitative methods comes the major problem: how to combine the two. Our present knowledge can't fully explain the "combination" method and principle, which have to be repeatedly operated and verified with the help of feedback information in the process of representation. Besides, qualitative thinking is often combined with figurative thinking, and it can be any one of the above discussed three thinking types while quantitative thinking largely uses logical thinking.

7.4.7 Another Thinking Methodology Used to Solve Specific Problems: Combination of Analysis and Synthesis (Also Taken as One of the Thinking Modes Solving Specific Problems)

"Analysis" is a general term for the thinking methods separating the whole into its important constituent parts to study the related motion mechanism. From the perspective of mastering things' motion mechanism the method of analysis is proper and necessary. Since things are universally connected and correlated, and such connections and correlations function differently to certain specific problems, analysis of the major relations is the necessary prerequisite for correct major results. But when going a step further, due to the dynamic change of motion and the complexity of correlations between factors, the results from analysis, having partiality for "reductionism", sometimes lack sound science and stresses too much on partial absolutory. "Synthesis" is a thinking mode combining components to form a connected whole which studies the trend and mechanism of the general changes taking into account various factors. An important thinking mode as it is, synthesis also has its possible defects: covering everything superficially but dealing with nothing in depth. So the combination of these two thinking modes with opposite characteristics according to the principle of mutually being both opposite and complementary can be scientific. But the scientificity depends on the degree of the organic composition of the combination in the light of specific problems.

We have discussed the general level on which problems are solved, and the basic thinking methods for specific problems; still there are more levels which can be extended deeper. E.g. the design and analysis method at the level of artificial system, which will be discussed in the following section, is in fact a method for the design of specific objects' systems, and a thinking method as well. The thinking method of different levels can be used mutually. Thinking methods, an induction and summarization of the thinking activities aiming at solving problems in the process of humans' survival and development, are in dynamic development and therefore know no bounds. The methods introduced above are an induction or summarization, and exemplifications of numerous thinking methods, aiming at generalizing the achievements of human thinking activities. But such great achievements as we've made, we should not stand still and refuse to make progress. So the following section will touch upon the computation science supported modeling and simulation method which is the modernized development of transformation method—the relation mapping inversion method—in the fields of science and technology and design, and etc. Social development and development of human thinking methods are a representation of the principle "survival of the fittest in the co-evolution of cosmos, large and small" in system theories.

7.5 Hierarchical Design Method

Humans have formed various design methods from constant thinking and summarization in conducting practical activities, among which hierarchical design method needs our special attention for it has to take account of many levels and the problems it deals with are complicated. More importantly, an outstanding scientific method can affect the success of the system as a whole. This section is dedicated to a brief introduction of several fundamental frame methods.

7.5.1 Modeling Analysis and Synthesis Via Human Brain's Abstracted Generalization

This method, when reviewed from the perspective of the thinking method for problem solution, is of the domain of "naturalization" method, in which problems are naturalized in human brain and there models are constructed via abstracted generalization. It is in the domain of models can methods for and routes to solutions be found.

This method is composed of two both independent and interrelated parts: the construction of models and the utilization of models to solve problems. After long process of evolution, human brain has possessed abstract generalization ability which is one of the most precious properties of human brain for if there's no abstract generalization ability there would be no complete thinking ability, nor the ability to solve complicated problems. If we go half a step further on the basis of this ability, we come to the construction of models. This is a process of abstract generalization, disentangling truth from falsehood (Falsehood here refers to the false appearance of things.), and is like "giving up a rook to save the king in chess", i.e., discarding the minor factors to grasp the major ones, based on which effective specialized and comparatively regular thinking procedure (including necessary accessories sections or thinking modules) on problems is formed, which is a constructed model.

Types of models:

Figurative thinking model (E.g., military commanders need construct figurative thinking model in overall operational planning.)

Mathematical model, which uses mathematical methods to represent models on the basis of logical thinking, e.g., using certain specific set in mathematical methods to represent and then solve problems.

Model in kind and model in semi-kind based on abstract generalization, mixed model in which different types of models work together, facing the object, to solve problems, and etc. The types of models follow no set form and are flexible; the key point is that they can grasp the essence of problems and precisely and successfully represent and solve problems in the end.

The purpose of constructing models is to help to solve problems and design systems, with which, it comes to the second part: the utilization of models. Computers are now widely used to analyze and synthesize problems to eventually find with high efficiency (in terms of manpower, material resources, and time) the basic routes or schemes for problem solution. So what distinguishing features does modeling via abstract generalization with the support of system theories and artificial system design philosophy—simulation analysis and synthesis have? The following four points are proposed temporarily.

In artificial systems motions are of multiple levels, sections and phases, so the construction of dynamic model systems varies on different levels and sections. Models of systems general level is the first to be constructed, then the model system can be gradually completed, in this process models of sub-levels should give feedbacks to and rectify models of general level, lower levels upper levels, and contents of models should be compatible with the environment in transverse direction, and extend to and interlink with lower levels.

In constructing models the properties of artificial systems should be paid special attention and responded correctly, e.g., systems function self-organization mechanism, concept of order parameter (among numerous factors only minority of them take effect), the dialectical unity of function and structure which is the basic elements for the survival of system, dual existence of the "acquired" and the "paid", and etc. Models should not miss or disobey these basic principles.

In terms of the method of modeling, we have to take into consideration the connecting and "assembling" of group models. The effective approximate simplified methods of quantitative analysis in system theories can be applied to the construction of basic models, e.g., random factors work jointly with factors with determinacy in constructing models; adiabatic method but with reduced independent variables and number of differential equations, and methods approximate to nonlinear equations (of more precise type of Fokker-Planck).

Due attention should be paid to the complete understanding and mastery of the core mechanism and results of the available models, based on which, to accumulate effective models standby application. Under the conditions that the mechanisms of new problems are same or similar to the available, we can use the available models for the solutions to new problems, which can cut down on energy and time, improve efficiency, and enable us to concentrate on the integration and application of models and modeling of the core new types of problems.

7.5.2 Relationship Mapping Inversion (RMI)

RMI is also of the "naturalization" method series, in which it also extends the modeling method a step further to specific problems. Yet of course RMI can solve problems of different levels and sections, not just applied to the general level of system.

RMI is a schematic method applied universally belonging to the category of general methodology of science, involving two steps to solve problems: relationship mapping and inversion.

Here is a case in point to illustrate this method: when one combs in front of a mirror, the "hair" in the mirror is actually the mapping of one's hair; combing in front of a mirror is a "relation structure" reflecting the relationship between the comb and hair; the practical effect—combing through the hair is the inversion of such relationship in the mirror. This can be better illustrated in Figure 7-1.

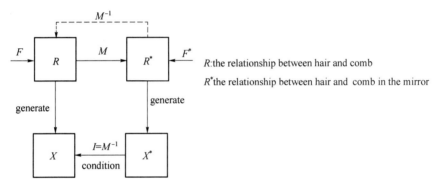

Figure 7-1　Relational mapping inversion method

Another case in point: we construct a model on problems for the purpose of using it to simulate and eventually solve problems; new models can also be constructed based on the above prototype model, i.e., R (the prototype) is via relationship mapping used to construct R^* (R^* model shall accurately reflect the essentials of R) in which effective measures are taken to obtain in the model domain the expected result x^*; the result is then reflected to R (including F, the inverse mapping of F^*), i.e., the due expected result x.

The above is only a simplified illustration. Things can be complicated, e.g., if in model R^* conditions are not enough to generate the expected x^*, partial structural relation c^* has to be added to obtain x^* (c^* can be added in various ways; it can also be a leap "insight".); then c^* is inversed to the prototype image domain to obtain c. This can be summarized as in Figure 7-2.

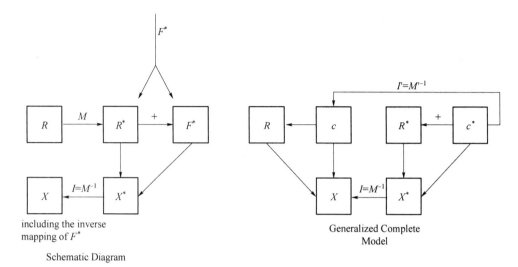

Figure 7-2 Schematic diagram and complete model

7.5.3 Hierarchy Analysis—a Method Combining Quantitative and Qualitative Methods Aiming at Solving the Problem of Decision-making Under Multi-Factor Environment in the Process of Design

This method generally follows the following steps:

Make clear the problems

Construct hierarchic structures

Seek single weight at the same level

Generalized check

Seek combination weight at the same level

Construct hierarchic structure

Problems are divided into several levels: the first level is that of the general target, the intermediate levels include the norm level, the specific target level, the constraint level, etc. and the lowest level the level of schemes to solve problems or the level of measures; under more complicated conditions, a feedback link is also an indispensable composition. See Figure 7-3.

For example:

Seek weight at the same level

To seek weight on the same level reflects the relative importance of the factors on

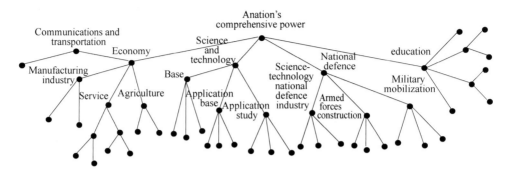

Figure 7-3 Comprehensive national strength stratification system

this level to certain single factor C in the higher level. It's also the base to calculate the combination weight of factors of each level on the general target, making one-to-one comparison of the factors of certain level according to certain norm and consequently constructing judgment matrix following the scale shown in Table 7-1.

Table 7-1 Suggested scale form

Scale	Importance	Notes
1	$A_i A_j$ share equal importance.	1) $A_i A_j$ are two comparisons on same level
3	A_i is of more importance than A_j.	judgments are based on certain norm (i.e. certain factor in the higher level).
5	A_i is of obviously more importance than A_j.	
7	A_i is of far more importance than A_j.	
9	A_i is of absolutely more importance than A_j.	

$n \times n$ matrix can be first obtained, we name it the judgment matrix $A = (a_{ij})$

As for A, we first calculate its largest characteristic root λ_{max}, then derive its correspondent standardized characteristic vector w, i.e., $Aw = \lambda_{max} w$.

The components of w we obtained here: $w_1 \ldots w_n$ is the importance of n factors, i.e., weight or weight coefficient; its largest characteristic root can be derived by methods of linear algebra or the approximation method. With the weight factors of each level, further merge and integration of them can be made to make decisions. In this process, a series of technical measures have to be taken, including consistency test, fragmentary data processing, which will not be elaborated here. The method of stratified analysis can also be applied to the analysis of index system of artificial systems to derive the relative importance and the implied relevance. In case that complete realization of the index is hard to achieve, this analysis is necessary.

Besides the decision method based on w, there is also another decision method available: "Gathered discussion room" proposed by Professor Qian Xuesen. Group discussions help to find new technical routes, and can attract scientific and technological personnel with different skills and knowledge as well, which is very important because nowadays scientific and technological achievements have been increasing in geometric progression yet in practice each individual's knowledge and specialties are not all-inclusive. Besides, successful design of a complex system often needs sciences and technologies of different disciplines working cooperatively and interwovenly. Therefore, a design group should include experts with interdisciplinary backgrounds.

7.5.4 Design Method Supported by Computation Science

Here this section will not repeat what's already discussed in Chapter Five about this method, but stress on its characteristics.

The system design method supported by computation science has the following obvious and unparalleled characteristics which render it very suitable for the research and development of complex systems.

1. The computation method and computation basis based on mathematics (combination of software and hardware, including data base and professional personnel factors) can represent and describe a complex relation set with numerous relations in it, e.g. the relation integration composed of up to hundreds of relations; the economy model of the Federal Reserve of the United States consists of 250 relational expressions.

2. The advance of computation science enables it to support model combinations of different types to dynamically represent from multiple sections and levels and describe in detail a complex motioning object according to its classes and divisions. That's why we say it is suitable for the research and development, design and analysis of complex systems.

3. The open structure of computation science draws as well as needs experts from different academic disciplines and with interdisciplinary backgrounds to form a team to do the work of model construction, simulation computation, and result analysis, feedback and rectification. Computation science can therefore be widely applied to various practical fields, we can say to no field it cannot be applied, and constantly accumulates and develops and gives feedbacks to application in the process of

application. It can therefore be inferred that computation science will further develop as an irreplaceable method for the design of complex systems, but it is worth pointing out: The development process into the aforementioned positive feedback loop is an important part of the process.

7.5.5 Method of Multi-Living Agent

See Chapter Five for detail. And this method will be further elaborated in Sections 7.6.3 and 7.9.

7.6 The Scalable Index System of an Artificial System

Human society's evolution makes more major artificial systems indispensable to social existence and development. To bring into full play their effects and incorporate them into larger systems for co-improvement, accurately representing them becomes necessary. Accurate representation of systems can be divided into many types and has different uses. In the design of artificial systems representation of systems appears as an index system which unifies the following opposite factors for the purpose of more accurate representation of artificial systems performance: analysis and synthesis, quantitative and qualitative methods, part and the whole, existence and conditions, necessity and possibility, reality and future, and etc. The index system works as the basis and criterion in the design (development), appraisal and evaluation, service, operation, maintenance and development of artificial systems.

7.6.1 Scalable Four-Dimension Comprehensive Index System

We propose an index system composed of performance dimension, economic cost dimension, temporal dimension, developability dimension, and their correspondent multi-level sub-dimensions, among which the performance dimension is the most essential one in terms of service but has to cooperate with other dimensions to form a complete dialectical index system in which the designed artificial systems can have the function of longer-term services. The index system has necessarily the quantitative description part, the subject described and limited by which can either be the direct quantitative limit to or the limit to the necessary conditions of the service function (if

Chapter 7 Major Points in the Design, Thinking, Methods and Process of Artificial Systems

only the two equal in value). See Table 7-2.

Table 7-2 Multi-level performance dimension index system

A performance dimension	Basic performance fractal dimension	System's general basic performance	
		Decomposed major basic performance 1	
		...	
		Decomposed major basic index n	
	Service performance fractal dimension	Service convenience	Service starting convenience
			Function switching convenience
			Storage convenience
		Service reliability	
		Service safety (in normal environment)	
		Maintenance performance	
		...	
	Counter safety performance fractal dimension in special counter environment	The first sub-dimension of counter performance	
		...	
		The nth sub-dimension of counter performance	

The national defense military systems, the operational equipment in particular, are originally intended for fighting and countermeasure and certainly have fine confrontation performance while civilian artificial systems such as fire fighting and rescue need also possess certain confrontation performance. In the case of larger yachts, they have to have the confrontation functions to confront sharks' attacks while maintaining undamaged.

Appreciation function fractal dimension

To arouse the user's keenness and appreciation of a product, the designer integrates the brand-name effect into its functions to endow its contours and human-machine interface with aesthetic value and human characteristics. See Table 7-3, Table 7-4, Table 7-5.

Table 7-3 Economic cost dimension index system

B economic cost dimension	Development cost sub-dimension	Cost in the pre-development stage	Information collecting and analysis (prior period planning cost)
			Application research (on the required advanced technology)
			Product pre-development (on the pre-research and development of the new product)
			...
		Project development cost	Project sample machine development
			Sample machine test experiment
			Preparation for batch production
			...
	Production, after-sale service and marketing cost fractal dimension	Production cost	Materials
			Production and manufacture
			Test
		After-sale service cost	
		Marketing cost	
	Use cost	Operation cost...	
		Maintenance cost	
		...	

Table 7-4 Time dimension index system

C time dimension	Predicted term of validity of functions	...	
	System's service term fractal dimension	Service term	
		Service calendar term (the total term in which the system can be used, regardless of work)	
		Maintenance period	
	Task finish response time fractal dimension	System task finish response time	
		Sub-task 1 finish response time	
		...	
		Sub-task n finish response time	
	Users' internship		
	System development period	...	
	Production period	...	

Table 7-5 Development factor dimension indicator system

D development factors dimension (which conducts analysis in the generalized space and the temporal domain on system's presentness while bearing its future development in mind.)	System operation and survival "bottleneck" factors fractal dimension	Property dimension major bottleneck analysis vs. overcome strategies	
		Eonomy dimension major bottleneck analysis vs. overcome strategies	
		Temporal dimension major bottleneck analysis vs. overcome strategies	
	Fractal dimension of opposition, environment and other factors influencing the system operation and survival	Indirect influence factors fractal dimension: e. g. the environment factor changes; products of the same kind emerge.	
		Direct influence factors fractal dimension: the emergence of countermeasures and technologies reduces or deprives the system service function.	

Among the development dimension factors those that seriously affect the system's operation and survival should be included in relevant index requirements, and room for development should be left for those that can be important to the future but lack accurate definite conditions at present in order to realize them in series development.

7.6.2 Connotations of the Index System

1) Relationship between the system's survival and the index system

The entire process starting from the emergence of an important artificial system through its service to mankind to its dying out can be summarized as the process of dynamic unity of a group of interwoven and opposite contradictions representing the existence of system. These contradictions are generally composed of multiple interwoven contradictions generated by system inner (between the system as a whole and the fractal system and the sub-systems) and outer operation environments and the user service. When they achieve conditional dynamic unity in motion they can represent the system existence service, and it's at this time that the properties of "condition" and "service" are completely defined by the index system. So when analyzed from a larger scale in the space-time domain, the index system actually refers to the improvement and "new emergence" of the corresponding system in dynamic changes.

2) Quantitative index connotations and the inclusive contradictions

The quantitative provisions of the index system compose the quantized relations (the quantized expression of the order parameter in the function mechanism operation, and of the sufficient and necessary condition for the full play of the function, and the mixed expression of the two) between specific conditions defining the system operation service. All provisions are generally vectors of certain sub-space in the space-time domain, whereas different provisions correspond with different sub-spaces (different coordinate systems), and they form mutual correlations for there is no orthogonal relations between them, which generates contradictions in design when trying to realize the requirements of the index provisions. For example, the difficulty in which the increase of the value of one provision will influence the decrease of another index. See Figure 7-4. For detail:

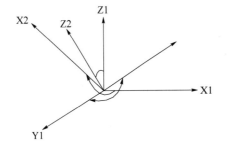

Suppose X1 Y1 Z1 is the coordinate system for provision A
X2 Y2 Z2 for condition B
If the spatial angle between X1 and X2>90°, then they play mutually anti-restrictive functions.

Figure 7-4 Interconnection and contradiction between quantitative indicators

The more complicated situation is that when the sub-space coordinate systems of the monomial index are not orthogonal, components of index vectors pin down each other in adjustment and thus difficulties are resulted in. It's concluded here that the complex correlations between indexes are one of the major reasons for the generation of difficulties in artificial systems' efforts to satisfy multiple index design.

3) Ideas on how to solve the contradictions included in the index system

Difficult as the formation of the index system framework and its quantitative definition are, man has never suspended his pursuit in developing artificial systems. We should summarize the deep ideas of the successful examples in man's pursuit and apply and spread their laws and advanced methods.

Use new scientific laws and technologies to specifically realize the philosophical principle of "the motion of Tao by contraries proceeds" ("Tao's" function is boundless and continuously develops towards the opposite, which thus forms a circulative development of unity of opposites which has no limits.) An artificial system's service existence is the unity of opposites achieved by system inner contradictions combining the environment; once the unity is destroyed, the system's existence in being will be broken and the system thus enters the state of turbulence; after the turbulence new development unity is formed, otherwise the system will die out. So when contradictions in properties happen in the emergence of a new system and it's hard to form unity, necessary new sciences and technologies are often introduced as new factors for the development of productivity to support the birth of the new system.

Use man's reason and wisdom to conduct in-depth analysis on the "required" and "possible" factors involved in the index system, and search when the two factors are in opposite states for the essentials for the realization of unity to promote them to achieve the state of unity of opposites; when the contradictions cannot achieve unity, often new sciences and technologies are employed to transform the strong opposite state into the unity of opposites. Only in this way, the contradictions are eventually solved and development is promoted.

As for the opposite contradictions of performance index and restraint condition index in the index system, we can transform their key physical quantity via the equivalence principle and mapping transformation principle to impel them to achieve the unity of opposites.

Use the conceptions of "integration" and "embedment" to form a new artificial system of "high gain" and "low pay" (But in high open conditions attentions should be given to the build-up effect of the safety weak points of all sub-systems for it can cause bigger safety "loopholes" of the integrated system.)

7.6.3 Characteristics of the Quantitative Index of the Index System

A. Universal characteristics

1) The quantitative index defines in generalized space domain, time domain, space-time domain, and variation domain. The domain of variation is the index quantitative definition of the variation rate of the variable quantity of the above domains in the same or interwoven domain.

2) Quantitative index must be in the definition domain and can be measured, defined and evaluated with definite accuracy.

B. conventional characteristics of quantitative definition

1) Define the existence scope of the quantitative index $X_{min} \leqslant \overline{X} \leqslant X_{max}$.

2) Define the existence lower limit of the quantitative index $X_{min} \leqslant \overline{X}$.

3) Define the existence upper limit of the quantitative index $\overline{X} \leqslant X_{max}$.

4) Use equal quantity to define the quantitative index $\overline{X} = X_0$.

5) The sequential definition of the quantitative index $\overline{X_1}\ \overline{X_2}\ \overline{X_3}\ ... \subset \overline{X_i}$.

Note: the property of $\overline{X}$ is not defined as single and can be multiple, e.g., it has the property of variation rate, and accumulation (integration), and etc.

C. Integrated quantitative definition

Regular integrated type

Strict integrated type has to satisfy simultaneously the definitions of multiple properties;

Optional integrated type selects one or several definitions from the definitions of multiple properties, for instance, the systematic selection of the configuration occasions.

Prerequisite transfer integrated type

Its logical structure is as follows: if ..., ..., then xxx, xxx.

A case in point is a tracking controller: when the motion velocity of the tracked subject $< V_{max}$, then the tracking error $\leqslant \varepsilon_N$;

When the motion velocity of the tracked subject $< V_{maxz}$, then the tracking error $\leqslant \varepsilon_{Nz}$.

7.6.4 Summary About the Index System

1) The index system is a brief yet complete representation of the system's survival and service, and survives or perishes together with the "system";

2) When the index system and the system reach the unity of opposites, they survive together;

3) On the premise that demands are satisfied, the index system which has "advance possibility" can promote the system's development, yet retard it in opposite conditions;

4) Scientifically and correctly fixing the system's index displays the scientific abilities, wisdom, and cooperation ideas and results of relevant people and teams.

7.7 Design Criteria

In the process of design, the following criteria are circling in all stages. In complicated situations indirect and covert violations of these criteria may take place, which should be guarded against.

7.7.1 Stage of the Formation of the Objective After the Perception of Contradictions

When in complex contradictory environments, special attention should be given to sorting out the principal contradiction perceived, the principal aspect of a contradiction and the conditions for the transformation of contradictions.

The formation of the objective of solving the contradictions via the design of a new artificial system and its operation and service is based on the definiteness of the conditions for the transformation of contradictions. The determination of the design objective is on the basis of the unity of the two opposite factors: "needs" and "possibility".

The essential points of the design objective, once determined, should not be altered in the process of design and development, for once altered, the design work has to start anew.

7.7.2 Stage of the Formation of the General Core Requirement for a New Artificial System

This stage is the start of the "thing" in which humans use the "object" (the new artificial system) to complete certain "objective", i.e., the transformation from "thing" to "object". Here the conditions for the survival of the "object" should be further ascertained, which involves two aspects: first, whether the service function of the artificial system can realize the service objective; second, since a new artificial system can survive in complicated environments only when satisfying certain constraints, relevant constraints have to be ascertained and satisfied as well.

The coverage of the service function of an artificial system must abide by the minimum complete axiom, i.e., satisfying the need of service completely but having no redundant functions (The realization of redundant functions has to pay a high price, and this consequently add to the additional conditions for survival.)

7.7.3 Stage of the Formation of Index System

The index system of an artificial system forms the basis for the design and testing and evaluation of the artificial system, so the definition of its service function and the set-up for the realization of the sufficient and necessary condition must observe the "minimal completion criterion".

Most of the index systems of artificial systems have complicated structures in which multiple interwoven contradictory relations are involved. So when setting up specific indexes for the index system, we should avoid the generation of contradictions in the mutual relations of the indexes, e.g., the increase in the requirement of one index is bound to affect the other index and causes its decrease; on the other hand, high requirements for both indexes at once will inevitably produce contradictions and eventually result in the failure in design.

7.7.4 Stage of the Formation of System Scheme

What's important in the formation of system scheme is the formation of multi-level and-section function self-organization mechanisms, and of the sufficient and necessary conditions for their existence. The self-organization mechanism of the system functions is in itself a whole and an organic integration as well of the function self-organization mechanisms of the fractal systems (sub-systems) (which can generate a "zooming" over the simple superimposition of the function mechanisms). So in the formation of system scheme the material existence of the above mechanisms must be ascertained.

The formation of the scheme must observe the minimal completion criterion.

The blind expansion of the application of new techniques should be prevented.

The requirements of the non-quantitative covert indexes should be met sufficiently, e.g. the important yet unable-to-be-described-precisely indexes in the developability dimension.

The formulation of the scheme should combine both the theoretical calculation and the practice to conduct simulation or analogous tests.

7.7.5 Stage of Technical Design

Observe the principle that blind expansion of the application of unnecessary new techniques should be prevented;

Observe the minimal completion criterion;

Implement the principle that the verification is conducted along with the design. The verification is of multiple levels.

Combine the fractional and decomposition design with the graded integration design to form a general integrated design of the system. In this process no omission or default is allowed.

7.7.6 Stage of the Test Manufacture and Testing of the Model Machine

The test manufacture of the model machine must strictly follow the design documents. The management of the out-of-tolerance workpieces beyond the stipulation of the documents should follow the standard procedure, and someone must be put in charge of this and keep a complete record of it.

In the testing stage, in the case that problems happen and alteration is necessary, the principle should be implemented that"Problems are solved without causing relevant side effects", i.e. the alteration can solve the problem found but will not produce opposite effects to other indexes' requirements, otherwise, it will result in complicated relevant alteration sequence.

All "corrections" must be included in the contents of "design solidification" so that they can be put into practice in production and service.

7.8 Major Steps and Contents of the Design (Development) of Artificial Systems

7.8.1 Co-Development of Human Society and Artificial Systems

1) Major concepts involved

Invention and utilization of tools evolve together with human society, this has become an eternal and important human development principle and embedded into the cultural field to become an indispensable composition of cultural development. In current society, a considerable number of artificial systems are actually advanced tools serving human society. Their utilization and development have promoted the evolution of the evolution mechanism of human society, e.g. the aerospace system helps to promote human evolution by enabling humans to carry activities in aerospace field. This is a new evolution mechanism; advanced medical equipment helps humans carry the medical work a step further from aiming at curing the existing diseases to a more advanced goal of ensuring health (the evolution of development mechanism in medical field). The development of artificial systems also consistently promotes new discoveries in science, and the invention and application of new technologies, and new research and design methods as well, which greatly improves man's cognitive and practical abilities. This once more proves that the development of advanced cultures is a major part in the evolution of human evolution mechanism. Here particular attention should be given to the important concept that the development of both artificial systems and the design of artificial systems (development) is a major part in the evolution of human evolution mechanism.

2) Major academic contents involved in the development of the design of artificial systems

The multi-level academic contents utilize the embedded design procedures to improve the design level, as shown in Table 7-6.

Table 7-6 The process of embedding multi-level academic content into the design

Basic theories		
Principle level	Philosophical level	Research and application of the law of the unity of opposites and the dialectic thinking principles and methods
		Research and application of the fields of the law about humans, the law of affairs, the law of matters and the law of beings
		The reflection of the law about humans and the law of affairs when related to philosophy of life, and the corresponding research and application
		...
	System theory	Application of dissipative self-organization theory
	Control theory	Application of self-organization theory and synergetic
	Complex theory	Non-linear control theory
		Research and application of complex theory and others
Basic thinking		Knowledge of thinking types and development of the corresponding application methods
Methods levels		Knowledge and application of universal solutions
Application basic level	Artificial systems design methodology	Artificial systems design philosophy
The application level		System's universal design theory and method: the multi-living agent design theory and method
		...
	Artificial intelligence theory and methodology	Characteristic model-based intelligent self-adaption control, etc.

7.8.2 Major Stages and Contents of Artificial Systems Design (Development)

The design of artificial systems generally follows the steps that the design work is carried successively from the top level to the lower, and the upper level design transfers its result to the lower and guides the lower to extend this "result". When an unsolvable contradiction takes place in the process of extending the design result in the lower level, feedbacks are given to the upper level and the design result of the upper level can be adjusted. This, only when circulated and repeated over and over again, can give birth to an artificial system with high performance and long service term. In practice, the completion of the design and development of an important new artificial system need the dialectically cooperation of at least three types of responsible personnel (team), i.e., those responsible for the development and management of this artificial system, those directly responsible for the completion of the design and manufacture, and the representative of the end-user. Though these three take different responsibilities in different stages, their dialectical unity is absolutely necessary.

This section involves a wide range of contents which, in practice, is an important yet complicated dynamic process of the activities of science and technology. Therefore it's hard to make it clear with shorter paragraphs. For further details please refer to 7.7, 7.8 and 7.9.

A. Stage of contradictions perception and of formation of the specific "objective" in solving contradictions based on the perception

The design work in this stage focuses on two parts: the correct perception of contradictions which is the prerequisite, and the subsequent, the formation of the specific "objective" in solving the contradictions. Since human society exists in contradictory movements, there are naturally various complicated contradictions from which humans select the important and solvable ones, and eventually solve them; this is the essential motive force for the design of new artificial systems. In practice, the process of analyzing in detail contradictions and selecting from them important ones that can be solved by designing new artificial systems is what we say "perception of contradictions", which is the general prerequisite for the design of artificial systems.

When "perception of contradictions" is finished, it's necessary to make definite the future result to be obtained by solving the contradictions; this is the "objective", which can be achieved via the service of artificial systems designed by humans. So the realization of the "objective" is a thing of future tense which has to abide by the guidance of the law about humans, the law of affairs, the law of matters and the law of beings, evolving from "being without form" to "being with form". It therefore is complicated and uncertain, and possesses challenges as well as opportunities. The personnel in charge of the management and development of artificial systems are responsible for the research and planning work of the "perception of contradictions and the formation of objective".

B. Stage of the formation of the general core requirements of a new artificial system under the guidance of the "objective"

The "objective" is formed to make clear the thing to be done and the result to be obtained. The subsequent work is to transform the thing to be done into an "artificial system", through which the thing can be done. So the realization of the transformation of an "objective" into design gives birth to a new artificial system which serves man, whereas as for the birth of a new artificial system the formulation of a general core requirement of a scientific system is an essential foundation and prerequisite for the smooth development of the follow-up design work. It is also possible to consider the "objective" as the platform and germ of the subsequent work.

A system's general requirements are actually the definition of the necessary and sufficient performance of the system by the general level of the system. Definition of an artificial system's general core requirements must be based on strict construction and analysis of the model and quantitative analysis of contradictions, only in that case, the unity solving the contradictions can be achieved. Besides, further analysis must be carried on to test whether the necessary and sufficient condition for the existence of the general core requirements is tenable. Only when the general core requirements and their necessary and sufficient conditions are both tenable, the work of this stage can be counted done for the time being. This stage still takes the personnel responsible for the management and development of artificial systems as the principal one, yet relevant professional teams can also join in responsible for negotiation and demonstration work.

C. Stage of the formation of the index system of the artificial system

The index system in this stage refers to the foundation and the criteria for evaluation and measurement based on which the entire design can be completed with the general core requirements as the prerequisite.

The index system includes the definition of the quantitative and qualitative description of the whole system's multi-level and section technical requirements, and the necessary and sufficient conditions that ensure the realization of the technical requirements.

As for the artificial system that is widely used, has relations relevant to multiple parties, and involves multiple even international manufacture and operation, its index system usually develops from being used only in specific special situations into criteria for the entire industry, the nation and even the world.

The work of the formation of the "index system" must involve those responsible for system development and management (including the consulting units) who take the principal part, and those who are responsible for the design and manufacture of the artificial system as well to achieve the unity of opposites.

D. Stage of the formation of the system scheme

The formation of the scheme is an important beginning of the practical design work after the personnel responsible for design discussed and confirmed the index system. So the correct formation of the scheme has far-reaching effect on the entire design.

The essential point in the design of the scheme is to determine the structure and composition of the entire system, the fractal system, the sub-system and the multi-level systems, and the self-organization mechanism of the artificial system's functions.

The design of the scheme involves not only technical design based on technical index, considerations should also be given to how the artificial system can harmoniously embed into the service environment to form living operation.

Good scheme design often has the vigor of bequeathability and extensibility in application. E.g., Boeing 737 has been extended to many types including 737—300, 737—400, 737—700, 737—800, and the production of these airliners had continued for more than ten years; another case in point is the sparrow-3 air-interception missile. Its CW semi-active missile guidance system is extended to ground-air missile and different countries have developed this system independently yet followed the same routine.

E. Stage of technical design

The work in this stage is the most specific and the heaviest workload for the personnel responsible for design. Whether it can be successfully completed directly affects the result of the design.

Technical design is in the charge of the personnel responsible for design. They will organize a high-level professional team in which the members share out the work and cooperate closely with each other.

A successful technical design is the unity of legacy and technical innovations, experience and practical solutions, and application of new methods.

Technical design consists of the general design for the entire system, fractal system design, and the design integrating fractal system and general functions.

F. Stage of the manufacture and testing of the model machine

In the design and development of modernized artificial systems, it's better to complete the design and development work with at most two rounds of model machine manufacture and test.

Process and manufacture the soft and hard wares of the components, sub-systems, fractal systems and the entire system based on the scheme design and technical design, and then test their functions; Conduct integrated test of the functions of components, sub-systems, fractal systems and the entire system, this is a rather complicated and arduous work, but has to be done by creating conditions and has to pass the integrated experiment on system functions, too.

7.8.3 Application of the Multi-Living Agent Method to Design

Since Chapter five has discussed this method and its theory, this section will focus on its application associating with the design steps.

A. Its application features

It decomposes the system into multiple living agents to facilitate in-depth analysis, and then to form fractal system living functions; and the comparatively strong integral function of the system is formed via the negotiation and coordination of the multiple agents of three levels (The coordination takes place between agents with the management and control agent as the leading force and on the condition that the management and control agent supports the administrator.) So the realization of the unity of the two opposite methods "analysis" and "synthesis" helps here.

The four-level (The agents themselves are included here.) adjustment and maintenance of the function livelihood render the system the function livelihood with stronger functions to adapt to the changes in environment and tasks, which further explains the concept of "process planning".

This method's feature of combining multi-level decomposition and analysis with multi-level synthesis and integration facilitates the construction of multi-level models to properly represent the system, and helps to obtain accurate design results by working together with the modelling simulation method supported by computation science.

The basic structure of this method can be embedded into and applied to all steps in the process of design, which makes the "design" more conform to principles, and consequently greatly improves the success rate of design.

B. Its application in design steps

See 7.10 for reference.

Stage of the formation of the "objective" after the perception of contradictions

The major features of the work in this stage are as follows: the objective agent which has function livelihood self-organization mechanism and works in given working environment is formed based on the core solution concept after the detail analysis of the contradictions (It's represented by $F^n(Q, I | R)$ as shown in 7.10.)

Stage of the formation of the system core requirement based on the preliminary materialization of the "objective"

Though this stage only involves the preliminary materialization, it is of great significance, for the formulation of a scientific core requirement is the essential base of the successful development of the subsequent work. The core requirement is the specific quantized representation resulted from the subtle quantitative analysis of the system function livelihood self-organization mechanism which is conducted based on the core solution concept and the practical environment. It is often a group of organized arrays. For details, please see the quantitative analysis of the components in $F(Q, T | R)$ and the core requirement formed by the quantitative representation of those conclusion numbers in 7.10.

Stage of the formation of the major index system

The core of this stage is to extend the core requirement of the system to generate the system property of function livelihood self-organization mechanism in working environment, and to construct corresponding index system whose solidified quantitative representation is the basis and foundation for the subsequent design work.

Stage of the formation of the system scheme

The core of this stage is to work out a specific system scheme with the functions defined by the index system and composed of multi-level multi-living agents by combining the system's universal structure consisting of the multi-living agents. This scheme requires the multi-level negotiation and coordination between multi-living agents to ensure the system's function livelihood in complicated environments.

Stage of technical design

This stage, based on the formation of the scheme, starts from the technical standard design of the system as a whole, e.g., the work flow of the system (including the regulations for negotiation and coordination between fractal system agents), various protocols, and etc., and then to the function livelihood design of the fractal systems and the agent level respectively. After the finish of the technical design of all agents are the design and dynamic test for the negotiation and coordination between multiple agents from the perspective of the whole system. The completion of the technical design requires such a cycle two or three times.

Stage of the manufacture and test of the model machine.

In this stage the application features of this method discussed above can be used to complete the work with higher efficiency.

7.9 Perception of the Off-normal States in Design and the Relevant Handling

The off-normal state in the design and development process strictly speaking includes two aspects: one refers to the abnormal state taking place directly in the process of design; the other is the off-normal state in a broad sense, referring to the possibility that the design being formulated at present can influence the normal service of the artificial system to the human beings. The right perception of the latter state is in fact a pre-perception which is difficult and the right handling following it is even more difficult. But in the design of some artificial systems, if severe abnormality of the latter state discussed above happens, prompt and decisive measures should be taken to avoid heavier losses. There is no lack of international precedents of handling problems of this kind. The root of this off-normal state does not lie in the detail, but the top factors in the design process, e. g. the purpose of design, and it's the result of the severe contradictions between the system core requirement and the dynamic and fast-changing service environment, e.g., the rapid deterioration of the indexes of the development dimension can influence the normal existence and service of the artificial system. In this case, effective measures must be taken to avoid more severe results.

7.9.1 Types of the Off-Normal State Generated Directly in the Design Process and Its Perception

A. Types and causes

1) When critical contradictions take place in the connotation of the design work in this stage, e.g., certain important index cannot be achieved, and contradictions are generated between the requirements of indexes;

2) When the sufficient and necessary conditions for the requirement of the index cannot be realized;

3) When the design connotation of this stage contradicts with the transfer requirement of the preceding stage;

4) When contradictions happen between the extended requirement of the design connotation of this stage and the design connotation of the following stage (level).

B. The perception principle

1) The principle combining the qualitative thinking and quantitative analysis.

All design stages must implement the principle that differential quantitative analysis and calculation must be based on the directional system qualitative thinking. Without the directional system qualitative thinking determining the theme and direction, the quantitative analysis and calculation based on modeling can often deviate from the right direction or be off the point, failing to grasp the key points, whereas only doing qualitative thinking can always result in too crude thinking and analysis. So in practice we should combine the two.

2) The principle in which contradictions influencing the "unity" take place.

Contradictions exist in all things. The existence of contradictions following the law of the unity of opposites forms the "being" of things. When a concrete "unity" cannot exist, the state of things will definitely change. All stages in the process of design have their concrete unity forms to represent the existence of "design", so when contradictions influencing the existence of design happen, the representation of design is in abnormal state. Attention should be given to this.

3) When one or two abnormal results happen in a battery of tests, their causes must be found and neglect of them is not allowed. The mechanisms generating such abnormal results must be detected and rectified, but the rectification on the other hand should not impose additional side effects.

C. Examples of perception of abnormal states in all stages

1) In the stage of the formation of objective via the perception of contradictions.

If in sorting out contradictions, the condition for the existence of the principal contradiction and the principal aspect of a contradiction are not clear and definite, the formation of the objective to solve the contradictions in this circumstance is risky due to the unstable "basis". For instance, if the sufficient and necessary conditions for the realization of the objective are not satisfied, though the formation of the objective is theoretically correct, it cannot be realized in practice.

2) In the stage of the formation of system core requirement.

The work in this stage is the extension and further specification of the preceding stage in which the objective is formed. If there are defaults in the preceding stage, the correct set up of the "system core requirement" will definitely be affected. The perception of this abnormal state mainly relies on the work of the objective formation stage, but not of this stage. But the contradictions between the setup of the system core requirement and the realization of the objective, as well as that between the realization of the system core requirement and the conditions for the realization should be done in this stage.

3) In the stage of setting the index system.

The focal point of the work in this stage is to perceive the contradictions between indexes formed by potential multi-level performance correlation. If the requirements of one index necessarily cause the decrease of the other index and eventually result in the failure in satisfying the requirements of the other, the potential abnormal state of this kind constructs latent dangers to the following multi-level projects, including system scheme design, technical design, test manufacture of the model machine, and etc.

4) In the stages of formulation of system scheme, technical design and the manufacture and testing of model machines.

If in these stages a few indexes are in critical values, or an important index doesn't meet the requirements in certain requirement condition, these can all be regarded as the abnormal state, the root causes of which should be analyzed and effective measures should be taken to rectify them. The environments in which the design, testing and evaluation are conducted are always simpler and more ideal than those in practice. So the abnormal states taking up in simple and ideal environments must not be ignored.

7.9.2 Perception of the Non-Normal States in a Broad Sense

The perception of the non-normal states in a broad sense is more complicated than the direct perception, for it cannot ignore in an open complex environment of a larger sphere the negative states formed by the combination of major external causes and the internal causes possessing the system properties. And when the effects of the perception of this kind can be seen is still in the future tense after the completion of the design of the artificial system. Due to the complexity, there is no mature and set measures to precisely perceive the non-normal states in a broad sense (Psychological analysis tells that precisely perceiving the future complex motion beforehand is unreachable.) But since this perception is of vital importance, people have been trying hard to realize it, largely relying on the integration of human professional knowledge and experience with the information obtained from outside to analyze the contradictions to make judgments,

Chapter 7   Major Points in the Design, Thinking, Methods and Process of Artificial Systems   195

which is of the human intelligent activities of deeper level based on the dialectical thinking.

7.9.3   General Principles for the Handling of Non-Normal States

A. In the implementation of the correction and remedial measures, we accentuate the principle of pertinence and promptness.

The correction and remedial measures should be directed against the non-abnormal states to improve the correction effect, whereas the promptness can reduce further losses brought by the lack of effective corrections in the design process. For example, the system scheme inadequacies cannot be rectified efficiently in the technical design level; the later the corrections to such inadequacies, the larger the losses. If the corrections are postponed to the batch production stage, tremendous losses will result.

B. The principle of using a practical and realistic approaches to equitably bear responsibilities for the mistakes occurred.

When mistakes occur in design and non-abnormal states take up, the top technical directors should not shift the responsibility onto others and blame the subordinates for no reason, for this is detrimental to the cooperation of the design team and doesn't help to find promptly effective measures to reduce losses.

C. The principle that the "upper level" stage bears responsibility for the mistakes and is responsible for the corrections as well.

The upper level stage (or the top stage) poses requirements (or restraints) to the following stages, and then inadequacies of the non-abnormal states result. In this case, naturally it's the "upper level" stage who is responsible for the inadequacies and has to rectify them. This principle plays fair and helps to rectify the inadequacies effectively and reduce losses.

D. The principle of switching the design "objective" to reduce losses in the case that the overall failure is inevitable.

When water is over the dam, and the overall failure is inevitable, we have to admit our failure as early as possible and switch the design objective to make the current design satisfy the new design objective. Then the current design can be transformed into a new design to perform new tasks, in this way a complete failure can be avoided and losses can be reduced. This is the last measure we have since we don't have any alternative. Here is a case in point. The "Iridium Satellite" project of the United States was originally a civilian satellite communication project, yet when failure occurred, the designers transformed it into a military satellite communication project and sold it to the military to reduce losses both in economy and reputation.

7.9.4 Brief Summary

The happening of the off-normal state in the process of design is what the design staffs don't want to see, especially when it's near the completion stage. But it's a possibility which cannot be absolutely avoided in the development of new things. So handling it in a practical and realistic way is of primary importance, only in this way can people promptly find scientific rectification and remedial measures. Only fundamental methods are discussed in this section, based on them scientific and technical personnel can bring into play their knowledge, experience and intelligence to find the most effective solutions in practice.

7.10 Examples: the Design of Air Terminal Defense Ground Environment (Based on Certain Foreign System)

The following sections will discuss by stages the research and development and design.

7.10.1 Stage of Formation of the Specific Objective of Solutions Via Perception of Focal Contradictions

The purpose of forming an objective (a future result) is to realize it, and the realization is done through completing certain thing with efforts. So from the formation to the realization of an objective is actually the process of handling affairs which certainly should directly conform to the law of affairs, and need the help of tools (matters) and to serve humans and the involvement of humans as well. So this process correlates with the law systems including the law of affairs, the law about humans, and the law of beings, with philosophy (including philosophies of all sub domains) as the core of the laws of deeper levels. This section, through analyzing the contradictory philosophy, discusses the objective formed and the core factors of and routes to its realization.

A. Perception of contradictions

Important ground facilities, including bridges, hubs of communications, converting stations of power plants, military top command posts, are the major attack subjects of the opponent. If aided by the air attack which possesses long range suddenness, and multiple effective attack measures, the opponent can destroy these important facilities with high efficiency. This constitutes a vital threat to our defense.

The serious threats to our defense by the air attack include:

All-weather full-time sudden attacks at any time

Diversified multi-batch and-height high-density attacks

Composite attack using multiple attack measures and with multi-level diversified attack goals and effects to realize the destruction with high efficiency, including missile attack outside the defense area, agile bomb attacks, anti-radiation missile attacks, electromagnetic attacks and conventional bomb and rocket bomb attacks.

With the knowledge of the characteristics and severity of the air attack, the solution to contradictions (the follow-up extension after the perception) can follow the principle of acting in a diametrically opposite way to achieve the state that the contradictions are both opposite and complementary to each other, i.e., we should actively destroy the opponent's attack destruction (This is an effective measure in a long run which reflects the process design.), not just accentuate our important facilities not being destroyed which is rather passive in itself. In this case, the principle of being both opposite and complementary to each other by acting in a diametrically opposite way conforms to the law of the unity of opposites, and is an important extension of the latter in the military countermeasure field. And it can be further extended to the military philosophy field to form the new principle of conserving oneself to wipe out the enemy whereas eliminating the enemy to preserve oneself, i.e., to achieve the goal of eliminating the enemy, we must conserve ourselves first. Conserving ourselves is the prerequisite for the elimination of the enemy, and the purpose of the elimination of the enemy is to conserve ourselves. These two are opposite yet reach a dialectical unity.

B. Formation of specified objective with multiple connotations

The objective, representing the integration of the matter to be done and the measures taken, has the following important connotations:

In major facilities air defense, only by constructing complex defense systems can we take effective measures to act in opposition to the enemy, including early warning system, regional defense system (implementing long range attack to the enemy delivery platform) and terminal defense system which is the last defense system that can conserve our forces and ensure the safety of our important facilities while eliminating the enemy. The terminal defense system is a new system; unlike the regional air defense system whose major aim is to destroy from long distance the opposite side's delivery platform and mother aircrafts, this system's core function is to destroy the destruction launched by the opposite side to our major facilities, for example, destroying their missiles, anti-radiation missiles, agile missiles, and mother aircrafts, too. Unlike regional air defense systems whose main objective is to destroy the opponent's delivery platforms and carrier aircraft by long-range strikes.

Under the opposite side's fierce attacks, the objective is fulfilled through destroying the opposite attacks. The terminal air defense system must have the ability of cracking down on the opposite attacks with high efficiency, accuracy and agility (in temporal domain) in the strong countermeasure environment, which can be expressed by $F(Q, T \mid \text{High Confrontation Winning Possibilities-R})$ in which R means the prerequisite of subduement in high countermeasure, T the agility index set, Q the accuracy effectiveness index set, and $F^n(Q, T \mid R)$ the high possibility index of destroying the enemy attacks composed of T and Q ... This is a centralized expression of the functionally active self-organization mechanism and a further expanded representation of S_0. Where Q is then decomposed in the meticulous design to meet the requirements of Q_1, Q_2, (systematic error up to the required accuracy matching, accidental error to begin with).

The terminal air defense system is the further realization of the high performance index goal R (Q, $T \mid$ High Confrontation Winning Conditions). We must fully utilize the integration of the effective information technology and the damage technology to lead to a terminal air defense system that combines the fire control system based on advanced information technology and the high-efficiency firepower damage system in the control of the fire control system, which actually is a specified objective extended a level deeper into the implementation level from the general objective level. This actually follows the thought that along the development direction of the ground air defense system the information technology and new damage technology are added to form some leaps in functions, and a new terminal sir defense system is thus formed.

The objective agent is formed with the support of academic factors, and then extended to be used in the follow-up work. Living objective agents represent the objective, and the beginning of the application of the multi-living agent method in the follow-up design and development work. The objective agent is formed through the following essential factors:

The planning that the objective is formed based on the perception of contradictions by a quality and capable person (team).

Conduct the follow-up work according to the principles of human design being and artificial system design being in design philosophy.

In the formation of this objective agent, the self-organization mechanism can be extended to living self-living mechanism, which is significant both theoretically and practically in that it functions in the strict restraint strong countermeasures environment. And combining them for the formation of "objective", materializing formation systems and requiring the extension of the principles to multi-active agent theories and methods.

The living self-organization mechanism can be expanded to form the multi-living agent theory and methods, on this account multi-level and living agents are formed and the systematic construction is completed, and the dynamic function required by the connotations of "objective" and the technical index can be fulfilled at multiple levels.

Major functions of the living objective agent are shown in Figure 7-5.

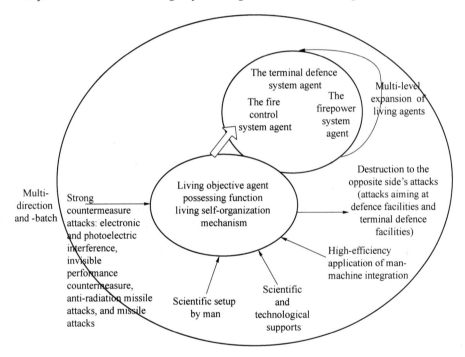

Figure 7-5 **Major functions of the living objective agent**

7.10.2 Initial Materialization of the Objective to Form the Core Requirement of the System

This section, on the basis of the qualitative functions of the living objective agent, uses the concept and methods of function living self-organization mechanism and multi-living agents to analyze and determine the quantitative system core requirements of the terminal air defense system. We adopt the quantitative index in system fulfilling tasks as the representation of the living degree of the living agents, then specify $F^n(Q, T|R)$ into terms $Q = Q(q_1, q_2 \ldots q_m)$ $T = T(t_1, t_2 \ldots t_n) T = T(t_1, t_2 \ldots t_n) R = R(r_1, r_2 \ldots r_r)$ to form the system core requirement F, and then to construct an analysis model with which we can determine the operations of the core parameters. It can also be considered as the analytical determination of the self-organization mechanism of the functional activity of the system layer of the objective agent and its sequential parameter set characterization.

A. Core system index

The following discusses the relevant terms in $F(Q, T|R)$ and the formation of F:

In $T = T(t_1 \ldots t)$, t_1 is the time used to search for and find the target, t_2 the determination and then the capture and tracing of the target, t_3 the control parameter formation time, and t_4 the time in which the firepower weapon takes aim and completes the preparation for firing. $T = t_1 + t_2 + t_3 + t_4$.

In $R = R(r_1, r_2, r_3)$, r_1 is the probability of success of anti-electronic interference in the search process of the fire control system, r_2 the composed probability of the success probabilities of both anti-electronic interference and anti-photoelectric interference in tracing, and r_3 the fire control system probability of destroying the anti-radiation missile attack, $R = r_1 \cdot r_2 \cdot r_3$.

In $Q = Q(q_1 q_2 q_3 q_4)$, q_1 is the reliable working probability of the fire control system, q_2 the reliable working probability of the firepower system, q_3 the defense system firing probability under multi-batch and-direction mass attacks, and q_4 the probability of destroying attacks in a single countermeasure firing, $Q = q_1 \cdot q_2 \cdot q_3 \cdot q_4$.

$F = (Q \cdot R|T)$, but since r_3 is of the same measure and model with $q_3 \cdot q_4$, in calculation only one term of them is taken into account to avoid repetition; the effect formed by r_2 is amalgamated into the density of the attack low (see the countermeasure analysis model for reference); besides, since there exist complicated relations between q_3 and T, when T is reserved beforehand, the value of q_3 will be influenced and thus the value of Q (This can be specified in the following section in which the analysis model of q_3 is constructed.).

Here is the determination of the system core index based on the association of needs and possibility in the light of relevant knowledge:

$$F(Q, T|R) = F(Q \cdot R|T<T_0)$$

$Q \cdot R = q_1 \cdot q_2 \cdot q_3 \cdot q_4 \cdot r_1 \cdot r_2 \cdot r_3$, in which the analysis and determination of q_3 is rather complicated and will be discussed in the following section. Here in this section we borrow the result to determine $Q \cdot R$, and other data can be directly determined according to relevant knowledge.

q_1 is the reliable working probability of the fire control system $= \dfrac{MTBF_{fire\ control}}{MTBF_{fire\ control} + MTTR_{firepower}}$

q_2 is the reliable working probability of the firepower system $= \dfrac{MTBF_{firepower}}{MTBF_{firepower} + MTTR_{firepower}}$

MTBF is the mean time between failures.

MTTR is the mean time between failure repairing.

q_3 is the defense system firing probability under multi-batch and-direction mass attacks. It's an important complex rigorous index which will be discussed in detail in the following section.

The performance value of q_4 can be obtained according to the cooperation between the chosen missiles, antiaircraft guns and the advanced fire control system.

r_1 is another important index. In system design, a comparatively satisfying result of it can be obtained by the adoption of multiple advanced technologies, selection from the design parameters including compression reaction time and correlated compression function distance, and the application of the selected parameters.

r_2 employs according to the fire control system numerous effective measures in the tracking phase and can achieve comparatively ideal results.

As for r_3, see q_4 for detail.

The following are the set values for $Q \cdot R = q_1 \cdot q_2 \cdot q_3 \cdot q_4 \cdot r_1 \cdot r_2 \cdot r_3$:

$q_1 = \dfrac{300}{300 + 0.5} = 0.998$, $q_2 = \dfrac{400}{400.5} = 0.999$, $q_3 = 0.9$ (to be analyzed in next Section), $q_4 = 0.92$ (The first interception for missiles, together with the second interception with accurate density firing in bursts within short range, can achieve high destruction probability.)

$r_1 = 0.90$ can be satisfied only when effective measures are taken to confront the other side's interferences.

$r_2 = 0.96$ is comparatively easy to satisfy when in tracking phase.

The reaction time of this system takes the value 6 seconds, which is of major advanced performance level. Analysis on it can be conducted in combination with q_4, and efforts should be taken to enable it to satisfy the agility reaction.

$F(Q, T|R) = F(Q, R|T) = F(Probability\ of\ Destruction = 0.71 | T_0 \leq 6s)$

The above results are not high with regard to the absolute value of the target destruction probability of the terminal defense system, but since they are the system comprehensive application effect indexes under the prerequisite of dealing with high-intensity attacks with a complex air defense system, it is still an advanced terminal defense system (or an advanced system with high livelihood) compared with the traditional one with a destruction probability of 0.3–0.4.

B. Analysis on the firing probability of major sub-index q_3

The following is the discussed countermeasure model to multi-batch continuous saturated attacks, presented with the queuing theory and the stochastic service system.

The enemy's multi-batch attack flow parallels the stochastic customer flow to be served, the air defense system the information desk serving the customers, and firing to destroy the enemy's attack flow the service.

When the fire control system is tracking the enemy aircraft or firing, the follow-up enemy aircraft can take the chance to attack our air defense front or fly over the security target. Correspondently, in stochastic service system, if the customer flow leaves over the information desk for receiving no service, we call it "defense breakthrough". So we can evaluate the countermeasure function of the air defense system via computing the defense breakthrough probability, the smaller the defense breakthrough probability, the stronger the system function. Schematic of defense breakthrough is shown in Figure 7-6.

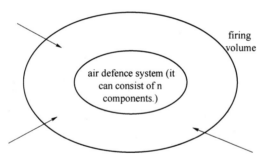

Figure 7-6 Schematic of defense breakthrough

Suppose the enemy's attack flow is stationary Poisson stochastic flow, the intensity of the flow is expressed by λ, its unit is batch/m, within the time interval τ, the Poisson flow's probability that k enemy aircraft arrive is $P_k(\tau) = \dfrac{(\lambda\tau)^k}{k!} e^{-\lambda\tau}$. When τ is very small, $P_2(\tau) = 0$, thus within short time intervals, the enemy aircraft have to arrive one by one (or group by group), its probability is $P(\tau) = \tau\lambda\, e^{-\lambda\tau}$; within the time interval τ, the probability that no enemy aircraft arrives is $P_0(\tau) = e^{-\lambda\tau}$. The average time of the entire process from contacting with a target, through intercept and tracking, to the end of firing is the average service time $1/M$, another major parameter is customers' average detention time $1/W$, which relates to the firing system's lethal distance, fire control system's discovery and tracking distance, enemy's aircraft's velocity, and our fragmentation weapons' flight velocity (These weapons include missiles and shells.).

Suppose the distribution of the two probabilities is of negative exponent, i.e.,

$$P(t_{service} < t) = 1 - e^{-Mt} \ldots 1/M \text{ the average service time}$$

$$P(t_{detention} < t) = 1 - e^{-Wt} \ldots 1/W \text{ the average detention time}$$

When customers cannot receive service within the detention time, they will leave (defense breakthrough).

Besides, there are two more important parameters needed in order to compute the firing probability:

$\alpha = \lambda/M$ multiplying the flow density and the average service time

$\beta = W/M$ the ratio between the average service time and the detention time

The following can be obtained from relevant references:

$$P_{firing} = 1 - P_{defense\ breakthrough}$$

$$P_{defense\ breakthrough} = \frac{\dfrac{\alpha^n}{n!} \cdot \sum_{s=1}^{\infty} \dfrac{s\alpha^s}{\prod_{m=1}^{s}(n + m\beta)}}{\sum_{k=0}^{n} \dfrac{\alpha^k}{k!} + \dfrac{\alpha^n}{n!} \cdot \sum_{s=1}^{\infty} \dfrac{\alpha^s}{\prod_{m=1}^{s} n + m\beta}} \cdot \dfrac{\alpha}{\beta}$$

In which n is the number of information desk, i.e. the number of air defense system. In the above expression $P_{defense\ breakthrough}$ relates to λ, n, α, β.

The larger λ is, the larger $P_{defense\ breakthrough}$;

The larger n is, the smaller $P_{defense\ breakthrough}$;

The smaller α, β are, the smaller $P_{defense\ breakthrough}$.

Smaller α, β can be achieved by reducing the service time and increasing the detention time. Reducing the average service time $1/M$ can simultaneously reduce α, β, this is high efficiency.

Thus we arrive at the important concept in the design of the fire control system, i.e. the core mechanism forming the fire control system's self-organization function: the reduction of the reaction time is more vital than the increase of the action distance; when the two contradict each other, first reduce the reaction time to make α, β take comparand and $P_{defense\ breakthrough}$ reduced to the desired value.

Further analysis tells that the service time (reaction time) mainly includes:

√ Time for hunting for and discovering the target

√ Interception time from discovering the target shifting to the tracking state

√ Holding time after firing: short antiaircraft guns can be moved immediately after firing, but missiles probably have to keep tracking after firing (e.g. irradiating target launching order)., yet missiles have higher precision and longer range of fire; shells can fire continuously and have low cost; advanced antiaircraft guns also have high precision and firing velocity, and are therefore suitable for the terminal defense.

√ In comprehensive computation, if $\lambda = 2\ batches/\min$, $\frac{1}{M} = 6s \therefore \alpha = \frac{6}{30} = 0.2$

$$\frac{1}{W} = \frac{2\ kilometers(\text{discover the target})}{700\ meters/\text{second}}, \beta = \frac{6}{30} = 0.2, P_{firing} = 0.94$$

If the attacking party develops, the vehicle attacks at Mach 1,000 (1,300 m/s),

$$\frac{1}{W} = 15.4s$$

So $\alpha = 0.2, \beta = 0.38, P_{firing} = 0.88$

Increase the search and detection distance to 28 km by the fire control system, compress the reaction time (service time) to 5 seconds by the fire control system, and above as a measure of functional activity maintenance adjustment (functional activity self-organizing mechanism development adjustment) any can adhere to $P_{firing} = 0.9$

If the attacking party's aircraft fly at a speed of Mach 1,000 skimming the sea at an altitude of 5 meters, the search radar must be placed on the mast of a 10-meter-high ship, high-precision data-guided fire control radar to intercept the target, the entire weapon system still maintains a 5-second reaction time, maintaining the probability of firing $P_{firing} = 0.9$, reflecting the application process of functional activity adjustment.

Then the attacking party to increase the attack density, and maintain the sea swept (5 meters) attack can be through the parallel two end defense system (n = 2), then P_{firing} can be significantly increased, $P_{firing\ n = 2}$ higher when the parallel two $P_{firing\ n = 1}$ to improve up to several times the huge, n to 3 or 4, to $P_{firing\ u = 1}$ base low to improve more obvious, but the composition of the weapon system is complex, the agent to increase the difficulty of coordination more, in the sea swept when the high-intensity attack, the Earth curvature affect the ship to discover the attack target, using the ship's helicopter radar to discover the target, can increase the target discovery distance as the basis of rapid response, at this time the helicopter and its radar additional data chain will be used as an active agent into the multi-active agent system composition.

7.10.3 Formation of the Major Index System

The formation of the major index system of the "system" is complicated, and the indexes necessary for applications are high. Besides, between indexes there are numerous intrinsic contradictions, which are intensified by the limitation of science and technology. So the index system formed in the first round through the compromise between "demand" and "possibility" need frequent adjustments based on feedbacks in the follow-up work.

A. Performance dimension

◇ Basic performance dimension

System's whole destruction probability $\geqslant 0.75$, $F(Q, T|R)$

System reaction time $\qquad \leqslant 6$ seconds

System searching target distance $\not< 20$ km RCS $1m^2$ target

$\qquad \not< 10$ km RCS $0.1m^2$ target

Searching angle resolution degree $2°$

Searching angle precision $\qquad$ angle $\not> 1°$

$\qquad$ distance $\not> 300$ meters

Target batch being tracked while in searching $\leqslant 10$ batches

Tracking precision angle $\leqslant 1$ milliradian longer waveband (angular velocity $\leqslant 20°/$second)

$\leqslant 0.5$ milliradian short waveband (angular velocity $\leqslant 20°/$second)

distance $\leqslant 10$ meters

Work frequency width $\not< 15\%$

The resolving and control precision of the future point of impact can be expressed by the general error as

$$\sqrt{\text{Mean square value of chance error} + \text{Mean square value of systematic error}}$$

The desired antiaircraft gun $\leqslant 08$ projectile radius of damage

missile $\leqslant 04$ missile radius of damage

The system reacts fast to the short-range firing with missiles having higher precision, antiaircraft guns' projectile high-velocity flight leading being small and able to form barrage to improve the damage ability.

◇ Service performance fractal dimension

MTBF $\not< 400$ hours (below secondary maintenance)

MTTR $\not> 0.5$ hours (below secondary maintenance)

1 operator

Construction time ≯ 15 minutes

Withdrawal and remove time ≯ 8 minutes

◇ Countermeasure fractal dimension

When the searching state is a weak link, the following are adopted:

Larger firing power

Wider frequency band, frequency agility

Narrower wave beam, low minor-lobe level

Advanced signal design and processing technology

The system reacts fast to reduce the searching distance.

Tracking state:

multiple tracking frequency range shunt option

Narrower tracking wave beam, low minor-lobe electrical level

Advanced signal design and processing technology

It's an effective countermeasure to smash the enemy's attack measures with accurate density firing (antiaircraft guns).

B. Economic cost dimension

development cost fractal dimension ≤ XXXXX ten thousand RMB

sales cost < XXXX ten thousand RMB in small batches

< 0.7 small batch (batch)

maintenance cost < XX ten thousand/year

others are omitted here.

C. Temporal factors dimension

lead time ≤ 5 years

system calendar ≮ 15 years

lifespan:

Performance work period ≮ 10 years (In this period since its service the system's performance does not fall behind and functions normally.)

production period ≤ 12 months

D. Developability dimension

Generally confidential, thus omitted here.

7.10.4 Stage of the Formation of the System Scheme

A. Overview

According to 7.9.2 and 7.9.3, theoretically the design team can start to form a system scheme composed of multi-living agents by combining the multiple agent theory and methods with their professional knowledge and experience. But in practice there often exist another level of contradictions between the mutual contradictions of the index systems of very high requirements and the technical realization abilities, and these contradictions complicate the determination of the system scheme and force it to make multiple rounds of operational selection which need the support from the computation science platform to make fast and convenient selection. This process, i.e., the modeling stimulation method supported by computation science is indispensably applied in the formation stage of the system scheme, which in application gives feedbacks to the results formed in the index system of the whole system for necessary negotiation and coordination.

In the system scheme the division and composition of the functions of multi-living agents are mainly based on the requirements of the performance and the developability dimensions, other indexes of the cost and temporal dimensions act as the restraint included in the negotiation and coordination process within and between agents in the formation of an advanced air terminal defense system scheme.

B. Composition of the multi-living agent system

After the formation of the system structure by multiple agents, attention has to be paid to the function negotiation-coordination between multi-living agents in the complex strong countermeasure environment, which actually is a continuous adjustment of the function living self-organization mechanism. For example, in the case of the system fast reaction characteristic ($\leqslant 6$ seconds), it needs the precise and accurate coordination between agents in the formation of the following firing process including searching for and discovering targets, selecting the target with the highest degree of danger, we identifying the target and track it, selecting a fire unit to transmit data and control signals to the fire control unit, tracking the present coordinate point, computing the point of impact, and the fire unit aiming at the point of impact to keep the time taken in the entire process $\leqslant 6$ seconds. This is only the negotiation and coordination between the most basic agents. Synthetically, the adjustment and maintenance of the system function livelihood, as discussed in Chapter 5, goes in four levels: the most basic level is the adjustment and maintenance of agents themselves; the second level is the function coordination between agents as shown in the above example; the third level is the coordination conducted to agents by management and control agents; the highest level is the function regulation and control to the entire system conducted by the administrator who makes decisions among the management and control agents. In fierce fighting, regulations and controls of multiple levels are often conducted simultaneously. For example, agents adjust their work modes to adapt to the environment to function, at the same time management and control agents provide to the administrator the target with the highest degree of danger, then the administrator decides on hitting the target. The administrator formulating new military tactics by making full use of the potential of the system and improving the performance of the system by applying new technologies working with the scientific and technological staff can be regarded as the function livelihood adjustment and maintenance of the system of deeper level, and the evolution of the multi-living agent system as well. See Figure 7-7.

7.10.5 Technical Design

In this phase, integrated designs between agents, sub-agents, and sub-subagents, and among multiple agents of multiple levels are conducted following the system scheme formed in the previous stage and the principle combining hardware and software, in which the core is the maintenance of the function self-organization livelihood. In specific designs we can use the existing modules to construct an advanced terminal air defense system, which involves a variety of disciplines such as information, control, machinery, mechanics, mathematics, computation science and technology, chemical engineering, materials, and management. The formulation of a system by a development and design team composed of scientific and technological personnel with different academic backgrounds by using the science and technology, materials and chemicals of different fields relates to a great many specifics, and here the details of the technical design are omitted.

7.10.6 Modulations Reflection of Living

Multilevel and much time of antagonism
Current time: Multi-living agents coordinate in current time, for example, anti-jamming measures of the radio Turn photoelectric
Short period: Tactical allocation, straight line, network support
Interim period: Single technology improvement
Long period: System structure forms basic unchanged, the agent structure improvement.

Chapter 7 Major Points in the Design, Thinking, Methods and Process of Artificial Systems 209

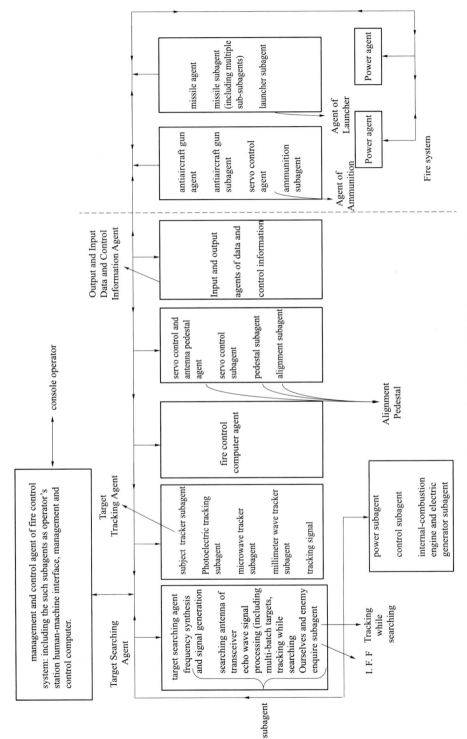

Figure 7-7 Composition of the multi-living agent system